Improving the Cybersecurity of U.S. Air Force Military Systems Throughout Their Life Cycles

Don Snyder, James D. Powers, Elizabeth Bodine-Baron, Bernard Fox, Lauren Kendrick, Michael H. Powell

RAND Project AIR FORCE

Prepared for the United States Air Force
Approved for public release; distribution unlimited

For more information on this publication, visit www.rand.org/t/RR1007

Library of Congress Control Number: 2015952790

ISBN: 978-0-8330-8900-7

Published by the RAND Corporation, Santa Monica, Calif.

© Copyright 2015 RAND Corporation

RAND® is a registered trademark.

Support RAND

Make a tax-deductible charitable contribution at
www.rand.org/giving/contribute

www.rand.org

Preface

Many new and legacy military systems rely on cyber capabilities to execute their missions. There are known threats to cyber aspects of these systems that create risks to the U.S. Air Force's ability to carry out operational missions. This report presents the results of a project commissioned by the Commander of the Air Force Life Cycle Management Center to develop policy, process governance, and organizational recommendations to improve mission assurance of acquired military systems throughout their life cycle in the face of advanced cyber threats. This research was completed as part of a project titled "Cyber Mission Assurance in Weapon System Acquisition." The work was conducted within the Resource Management Program of RAND Project AIR FORCE, and it should be of interest to the cybersecurity and acquisition communities.

RAND Project AIR FORCE

RAND Project AIR FORCE (PAF), a division of the RAND Corporation, is the U.S. Air Force's federally funded research and development center for studies and analyses. PAF provides the Air Force with independent analyses of policy alternatives affecting the development, employment, combat readiness, and support of current and future air, space, and cyber forces. Research is conducted in four programs: Force Modernization and Employment; Manpower, Personnel, and Training; Resource Management; and Strategy and Doctrine. The research reported here was prepared under contract FA7014-06-C-0001.

Additional information about PAF is available on our website:
http://www.rand.org/paf/

This report documents work originally shared with the U.S. Air Force in August 2014. The draft report, issued on September 9, 2014, was reviewed by formal peer reviewers and U.S. Air Force subject-matter experts.

Contents

Figures

Summary

There is increasing concern that U.S. Air Force systems containing information technology are vulnerable to intelligence exploitation and offensive attack through cyberspace. In this report, we analyze how the acquisition/life-cycle management community can improve cybersecurity throughout the life cycle of military systems. We focus primarily on the subset of procured systems for which the Air Force has some control over design, architectures, protocols, and interfaces (e.g., weapon systems, platform information technology), as opposed to commercial, off-the-shelf information technology and business systems.

We begin by identifying some fundamental principles of sound cybersecurity management and comparing them with the current state of cybersecurity laws and policies. These principles are drawn from the attributes of cybersecurity itself, findings from organizational theory, and elements of process control.

The desired outcomes of cybersecurity management are to limit adversary intelligence exploitation through cyberspace to an acceptable level and to maintain an acceptable operational functionality (survivability) even when attacked offensively through cyberspace. What constitutes an acceptable level of risk is determined by mission assurance risk acceptance. These outcomes need to be achieved continuously throughout the life cycle of a military system, from research and development through disposal. All phases are important, but the development and sustainment stages are particularly critical—the former because design decisions are made that can limit options in the future, and the latter because most systems reside in sustainment for most of their life cycle.

Managing cybersecurity risk has three components: (1) minimizing vulnerabilities to systems, (2) understanding the threat to those systems, and (3) minimizing the impact to operational missions. Multiple layers contribute to mitigating vulnerabilities: defensive measures to deny access to the systems, backed up by a robust and resilient design so that when attacked, the system degrades gracefully and recovers rapidly to an acceptable level of functional performance. The overall operational risk reduction will come from a combination of system security engineering,[1] assessment of how mission assurance is affected, and, because the cybersecurity environment is rapidly changing, adaptive solutions.

---

[1] We use the definition of *system security engineering* in Department of Defense Handbook MIL-HDBK-1785, *System Security Engineering Program Management Requirements*, 1995, p. 5:

> an element of system engineering that applies scientific and engineering principles to identify security vulnerabilities and minimize or contain risks associated with these vulnerabilities. It uses mathematical, physical, and related scientific disciplines, and the principles and methods of engineering design and analysis to specify, predict, and evaluate the vulnerability of the system to security threats.

The cybersecurity environment is inherently dynamic and complex. The literature suggests that well-managed organizations cope with such environments by choosing organizational designs that favor solutions obtained through decentralized coordination and collaboration of workers over those prescribed by standardized and formalized controls.

To manage cybersecurity effectively and to hold actors appropriately accountable, outcome-based feedback that measures the actual state of cybersecurity and its mission impact is needed, not just compliance with directives. An enterprise that only monitors and enforces compliance with its directives cannot effectively control the actual state of cybersecurity, and it telegraphs the message to the organization's members that compliance, not performance outcomes, is the enterprise objective.

Comparing these management principles with a detailed review of the laws and policies governing Air Force cybersecurity reveals a number of gaps and conflicts that impede the Air Force's efforts. Through this assessment, we arrive at four causal findings of concern, each of which has further consequences for security.

Our first finding is that the cybersecurity environment is complex, rapidly changing, and difficult to predict, but the policies governing cybersecurity are better suited to simple, stable, and predictable environments, leading to significant gaps in cybersecurity management. This trait has four consequences of concern:

- It prescribes solutions for military system cybersecurity in the form of controls that are not as comprehensive for providing security as sound system security engineering.
- The processes and security controls were developed principally with information technology systems in mind, not military systems, and hence solutions are not well tailored.
- The strategic goal of mission assurance is diminished in favor of tactical security controls.
- In relying on standardized and formalized security controls as the means by which cybersecurity is to be accomplished, the policy telegraphs to the enterprise that the implicit goal of cybersecurity is compliance with security controls.

Our second finding is that the implementation of cybersecurity is not continuously vigilant throughout the life cycle of a military system, but instead is triggered by acquisition events, mostly during procurement, resulting in incomplete coverage of cybersecurity issues by policy. This trait has four consequences of concern:

- These programmatic events come late in the design process, and therefore have little leverage to influence some critical design decisions that affect cybersecurity.
- Systems in programs beyond procurement, being sustained or disposed, get diminished attention relative to those in procurement.
- This policy structure leads to imbalanced risk assessment by favoring system vulnerability assessments over mission impact and threat.

- Management, oversight, and budgeting within the U.S. Department of Defense (DoD) are strongly structured around programs, whereas cybersecurity vulnerabilities cross program boundaries, creating a misalignment between the challenge and its management.

Our third finding is that control of and accountability for military system cybersecurity are spread over numerous organizations and are poorly integrated, resulting in diminished accountability and diminished unity of command and control for cybersecurity. These overlapping roles, and particularly the presence of a cybersecurity-focused *authorizing official*,[2] create ambiguities in decision authority and accountability. It is, for example, unclear who can make the final decision regarding risk to a mission: the commander or the authorizing official? And should a cybersecurity incident occur, who is ultimately to be held accountable: the program manager, the authorizing official, or the operational commander?

Our fourth finding is that monitoring and feedback for cybersecurity is incomplete, uncoordinated, and insufficient for effective decisionmaking or accountability. This trait has three consequences of concern:

- There are critical gaps in this feedback: It does not capture all systems, does not probe the consequences of cybersecurity shortfalls, and is not produced in a form that informs effective decisionmaking.
- The lack of comprehensive program- or system-oriented feedback on cybersecurity and the impact of cybersecurity on operational missions stands in stark contrast to the abundance of feedback on cost and schedule performance. This imbalance creates an incentive structure for program managers and program executive officers that favors emphasis on cost and schedule over performance, specifically cybersecurity.
- These deficiencies in feedback on cybersecurity inhibit individual accountability.

Two underlying themes carry through these findings: that cybersecurity risk management does not adequately capture the impact to operational missions and that cybersecurity is mainly added onto systems, not designed in. As partial redress for these findings of concern, we offer the following 12 recommendations:

1. Define cybersecurity goals for military systems within the Air Force around desired outcomes, while remaining consistent with DoD issuances.
2. Realign functional roles and responsibilities for cybersecurity risk assessment around a balance of system vulnerability, threat, and operational mission impact and empower the authorizing official to integrate and adjudicate among stakeholders.
3. Assign authorizing officials a portfolio of systems and ensure that all systems comprehensively fall under some authorizing official throughout their life cycles.

---

[2] An *authorizing official* is an official empowered to make risk-based decisions regarding the acceptance of cybersecurity risk of operating systems in DoD that process, store, or transmit information. In this capacity, authorizing officials report to the chief information officer (CIO) and assist the CIO in carrying out his statutory responsibilities. Authorizing officials replace the former designated approving or accrediting authorities. See Department of Defense Instruction 8500.01, 2014.

4. Adopt, within the Air Force, policy that encourages program offices to supplement the required security controls with more comprehensive cybersecurity measures, including sound system security engineering.

5. Foster innovation and adaptation in cybersecurity by decentralizing in any new Air Force policy how system security engineering is implemented within individual programs.

6. Reduce the overall complexity of the cybersecurity problem by explicitly assessing the cybersecurity risk/functional benefit trade-off for all interconnections of military systems in cyberspace, thereby reducing the number of interconnections by reversing the default culture of connecting systems whenever possible.

7. Create a group of experts in cybersecurity that can be matrixed as needed within the life-cycle community, making resources available to small programs and those in sustainment.

8. Establish an enterprise-directed prioritization for assessing and addressing cybersecurity issues in legacy systems.

9. Close feedback gaps and increase the visibility of cybersecurity by producing a regular, continuous assessment summarizing the state of cybersecurity for every program in the Air Force and holding program managers accountable for a response to issues.

10. Create cybersecurity red teams within the Air Force that are dedicated to acquisition/life-cycle management.

11. Hold individuals accountable for infractions of cybersecurity policies.

12. Develop mission thread data to support program managers and authorizing officials in assessing acceptable risks to missions caused by cybersecurity deficiencies in systems and programs.

We acknowledge that these recommendations, even if fully implemented, would not completely solve the challenges of cybersecurity. Further, some of these policies would necessarily require additional resources and a suitably skilled workforce to carry out the responsibilities, commitments that are difficult to make in a constrained fiscal environment. The fact is that there are no quick or easy fixes for achieving world-class cybersecurity. However, by adopting these recommendations, the Air Force would take a large step toward more effective cybersecurity of its military systems throughout their life cycles.

Acknowledgments

We thank Lt Gen Clyde D. Moore II for conceiving and commissioning this project. The work could not have been done without the continuing support of Kevin Stamey and Tim Rudolph. To understand the perspectives of the full range of stakeholders in the Air Force and other agencies, we met with and received help from scores of individuals in the research and development, acquisition, sustainment, intelligence and counterintelligence, test, and operational communities. They are too many to acknowledge individually, but we are very appreciative of their time and frank comments.

We were impressed by and thankful for the dozens of individuals we met who are deeply concerned about cybersecurity and who are passionate about doing it better. They are to some extent hampered by well-intentioned policies that limit their integration and coordination, and hence effectiveness. We hope that the findings and recommendations in this report will help facilitate their efforts.

At RAND, we thank Mahyar Amouzegar, Natalie Crawford, Ryan Henry, Myron Hura, Richard Mesic, and Lara Schmidt for helpful discussions. We especially thank Igor Mikolic-Torreira and Caolionn O'Connell for constructive reviews.

That we received help and insights from those acknowledged above should not be taken to imply that they concur with the views expressed in this report. We alone are responsible for the content, including any errors or oversights.

1. Cybersecurity Management

> "When the government purchases products or services with inadequate in-built cybersecurity, the risks persist throughout the lifespan of the item purchased."
>
> – Frank Kendall and Daniel M. Tangherlini[1]

Introduction

There are many goals in military system acquisition and life-cycle management, including keeping costs affordable and meeting milestones in a timely manner, but arguably the most important is that the military system performs the functions necessary to carry out military operations.[2] One of the goals of any adversary is to prevent military systems from doing so, or, failing that, to impede their operations with countermeasures.

A relatively new domain for challenging military systems is cyberspace.[3] Most modern military systems are so intimately intertwined with cyberspace that they depend on it for their fundamental operations. Many are further connected, either directly or indirectly, to other military systems, forming a complex system of systems whose capabilities are interdependent. The extensive dependence on cyberspace and the networking of military systems creates many enhanced and synergistic capabilities, but also leads to numerous potential vulnerabilities to military systems and operations from adversary intelligence exploitation and attacks through cyberspace.

Adversaries can potentially use the domain of cyberspace in various ways to challenge military systems. They can collect intelligence on these systems to steal technology and accelerate their own capabilities. They can use this intelligence to develop countermeasures. And they can use cyberspace as a means to directly attack U.S. military systems. Together, these pose significant risks to operational mission accomplishment.

Ensuring that military systems and their underlying information systems operate with acceptably low risk and adequately support operational missions in the face of existing and potential future advanced cyber threats lies in the realm of cybersecurity. But what does it mean for a military system to be cybersecure? *Cybersecurity* is defined by the U.S. Department of Defense (DoD) as the "prevention of damage to, protection of, and restoration of computers,

---

[1] Kendall and Tangherlini, 2013, p. 6.

[2] *Life-cycle management* is "the implementation, management, and oversight, by the [program manager], of all activities associated with the acquisition, development, production, fielding, sustaining, and disposal of a DOD system" (*Manual for the Operation of the Joint Capabilities Integration and Development System*, 2012, p. B-E-2).

[3] *Cyberspace* is defined as "a global domain within the information environment consisting of the interdependent network of information technology infrastructures and resident data, including the Internet, telecommunications networks, computer systems, and embedded processors and controllers" (Joint Publication 1-02, 2014).

electronic communications systems, electronic communications services, wire communication, and electronic communication, including information contained therein, to ensure its availability, integrity, authentication, confidentiality, and nonrepudiation."[4] This definition captures most of what people refer to as cybersecurity, but we adopt a simpler, more goal-oriented definition here, so that we can apply it clearly and usefully to military systems and information systems alike.

For the purposes of this report, we use the term *cybersecurity* to mean limiting adversary intelligence exploitation to an acceptable level and ensuring an acceptable level of operational functionality (survivability) even when attacked offensively through cyberspace. What constitutes an acceptable level of system risk, we argue later in this chapter, is determined by mission assurance management decisions. In short, this objective-oriented definition of cybersecurity means having effective counter cyber exploitation methods and survivability from attacks through cyberspace. The latter includes what is sometimes called *cyber resiliency* but is somewhat broader.

Cyberspace is quite a large domain. Systems that contain cyber elements span a range from those that are government-designed and have architectures, protocols, and interfaces over which the government has some degree of control, to those that are procured more as commodities, such as commercial products and services with established designs, architectures, protocols, and interfaces. Weapon systems and platform information technology (PIT) systems lie toward the former category, and business and communications information technology systems lie toward the latter. Our emphasis in this report is on the former category of systems. This category is broader than weapon systems and, although it applies to all PIT systems, what falls into the category of PIT is a little ambiguous and can change over time as definitions in policy shift. To avoid confusion, we are focusing our discussion on systems the government procures and for which it has some control over design, architectures, protocols, and interfaces, whatever these systems might be called in current policy. We will call these *military systems*. The work was commissioned by the Air Force Life Cycle Management Center (AFLCMC), and so the emphasis is on military systems managed by that center, but most of the discussion is also applicable to systems under the purview of the Space and Missile Systems Center and the Air Force Nuclear Weapons Center.

The acquisition/life-cycle management community plays a significant role in cybersecurity for military systems. In this report, we examine how to improve cybersecurity for military systems throughout their life cycles, including the phases of research and development, procurement, test, and sustainment. This report does not address specific technical solutions or proposals for new security controls. Rather, we address how policy changes, organizational roles and responsibilities, and monitoring and feedback can be adjusted to provide more coherent and integrated cybersecurity management.

---

[4] Department of Defense Instruction 8500.01, 2014.

This first chapter describes the goals of effective cybersecurity. Chapter Two describes the current state of cybersecurity laws and policies, particularly those related to acquisition and life-cycle management of military systems. The final chapter discusses shortfalls between the status quo of cybersecurity and what it ought to be in the form of four core, root-causal findings that have deeper consequences. We then offer recommendations for mitigating these shortfalls. In the remaining part of this opening chapter, we draw on principles of systems engineering, organizational design, and managerial process control to guide how cybersecurity might be better managed to achieve desired outcomes via integrated and coordinated efforts across the many stakeholders.

What Should Cybersecurity in Acquisition Achieve?

Counter Cyber Exploitation

Cyber exploitation of a military system is the extraction of information from or about that system by an adversary. Exploitation can be done by penetrating and extracting information from Air Force or contractor databases, and it can be done by accessing the military system itself. An adversary can use this information for a range of purposes, including stealing technology, assessing Air Force capabilities, developing countermeasures to Air Force systems, and preparing intelligence for an offensive attack against the military system.

For cyber exploitation to be successful, an adversary needs to gain access to useful information and exfiltrate that information before being detected and blocked. Access can be gained through software infiltration or by implants in firmware or hardware introduced through the supply chain. These access points can, in principle, be indirect via a less critical system that has some information exchange with the system containing the more critical information. Hence, to have effective counter cyber exploitation, the system owner needs to identify the most critical information, and the adversary needs to be denied access to that information; if the adversary gains access to critical information, he must be detected and blocked from successfully exfiltrating it.[5]

Counter Offensive Cyber Operations

Offensive cyber operations aim to do harm to a military system (or via a military system) by attacking through cyberspace. An offensive attack against a military system can, in theory, do any harm that can be inflicted by the software controls of the system. Damage can come in the form of the owner partially or fully losing the ability to use the system, the loss of active control of the military system, or the adversary taking full control of the system and using it according to

---

[5] Deception can also be used, in its simplest form, by luring the adversary to false information and allowing the exfiltration of that false information.

his own ends. The latter could result in using control to cause self-destruction or attack personnel or another system.

To successfully attack ware (soft, firm, and hard) through cyberspace, an adversary needs access to the ware, a flaw to exploit that affects the ability to carry out the mission, and the capability to do so. The capability to do so generally entails deep knowledge of the military system being attacked and how it functions. It is probably a fair estimate that the more consequential the offensive cyber attack, the more knowledge the adversary tends to need. Some of that knowledge will generally need to come from effective intelligence collection, including cyber exploitation (thus linking counter cyber exploitation and counter offensive cyber operations).

Countering offensive attacks through cyberspace has two facets. First, preventing attack requires limiting access, limiting flaws whose exploitation could significantly affect an operational mission (e.g., software assurance), and limiting the ability of an adversary to learn about U.S. systems and missions. Goals and methods of this facet overlap considerably with the goals and methods of counter cyber exploitation. Second, operating at an acceptable level of functionality after an attack occurs requires a military system design that can absorb and recover from an attack. The ability of a system to absorb an attack but still function at some acceptable level is often called *robustness*. The subsequent ability to recover from an attack by restoring either partial or full mission functionality is often called *resiliency*.[6] Cybersecurity against offensive cyber attacks, then, is layered.

The first layer is defense. If the defense is breached and an attack occurs, the attack is met by a robust military system, and whatever degradation in function that military system suffers, it recovers quickly by being resilient. (This layered solution applies at both the military system and mission levels.) These ends can be achieved by combinations of redundancy, diversity, dynamic adaptation, replacement of components when they fail, surgical deletion of infected components to stop contagion, architectural considerations for robustness and resiliency, sharing of resources with the adversary to deter attack, and protection/defensive measures. At the military-system level, like the case of counter cyber exploitation, the defensive layer is achieved by integrated security engineering of these techniques in the design phase. Robust and resilient military systems arise from strong systems engineering and an understanding of how humans interact with those systems, all guided by functional performance requirements.

Functional performance requirements for robust and resilient systems differ from those aimed at defense, such as security controls placed on the perimeter of a system to deny access. Those for defense contribute only to defense and depend on the threat environment. Requirements for robustness and resiliency are not reactive or outward-looking—they do not emerge from the threats that adversaries pose. They are proactive and emerge from the need for

---

[6] The terms *robust* and *resilient* have no universally accepted meanings in the Air Force, but are both widely used in this context. Our definitions are broadly consistent with other usages.

continuing functionality of the system. Cyber robustness and cyber resiliency are in this sense no different from other survivability requirements—they are outcomes based on operational requirements and not on regulatory standards. They are agnostic of how detrimental effects might befall a military system and do not rely on intelligence estimates of evolving threats. No direct need exists for specifying evolving threat modes, only that the system has, for example, a requirement for a certain level of functionality in the face of events such as the loss of data integrity, failure of a processor, or failure of an entire computer system.

Managing Cybersecurity Risk

Cybersecurity risk to an operational mission is defined as the product of three components: (1) vulnerabilities of systems, (2) threats to those systems, and (3) the impact to operational missions if those threats exploit system vulnerabilities. Cybersecurity management of the risks of cyber exploitation and cyber attack requires a balance and integration among all three components of risk. The management of each of these three components of risk is somewhat different.

The goal of minimizing vulnerabilities to systems by identifying potential vulnerabilities and their mitigations—in the form of denying access and having robust and resilient designs—is the primary responsibility of the life-cycle management community, and the program manager in particular. The goal of minimizing adverse operational mission impact—or, said another way, ensuring operational mission accomplishment—is the responsibility of the mission owner, often a Core Function Lead Integrator (CFLI). These two goals occasionally come into conflict and create tension in cybersecurity management. Policies and organizational constructs need to recognize and resolve this tension.

This tension arises from two sources. First, there are numerous instances in which cybersecurity can be achieved to some degree by either changing the system or changing how the system is used. Overcoming poor operational practices can require quite elaborate (and sometimes quite expensive) design solutions, if it is even possible to completely compensate. Likewise, poor design can require stringent operational practice, and, even then, the military system might remain vulnerable and the missions it supports at risk. Which solution to pursue, or what combination of solutions, is a decision that spans the stakeholders of the life-cycle management community and mission owners.

Second, how critical a system vulnerability might be depends on how that system is used to support operational missions and which missions it supports. Missions are accomplished by a combination of doctrine, organization, training, materiel, leadership and education, personnel, and facilities, together with policy (DOTMLPF-P), and missions comprise multiple systems. Mission assurance is not the same as each system maximizing its cybersecurity. For some systems, more risk will be accepted, either because that system is less critical than others to operational missions or because its potential vulnerabilities are less than another system's. To varying extents, shortfalls in one area of DOTMLPF-P can be compensated for by another.

Information for making decisions regarding accepting *mission* risk extends beyond the purview of a military system's program office, and effective cybersecurity management requires careful coordination and integration of efforts among multiple stakeholders.[7]

Finally, the assessment of system vulnerabilities and mission impact must be done in light of the threat environment. The threat information needs to flow continuously to stakeholders to support decisions throughout the life cycle of the military system. Some of that information early in the development phase will be more general and less certain. As the system design matures, is fielded, and eventually resides in the sustainment phase, information needs to be more specific and timely so that countermeasures can be effectively employed.

Challenges for Managing Cybersecurity

What do these goals for cybersecurity imply for the management of cybersecurity? In this section, we first discuss challenges posed by inherent attributes of cybersecurity before moving on to general management challenges faced by any endeavor. We close with a section outlining a proposed framework of general principles for sound cybersecurity management, which will serve as a baseline for comparison with the status quo.

Attributes of Cybersecurity That Inform How It Needs to Be Managed

First, the challenges in implementing effective cybersecurity are technical and involve attributes of systems that are integral to their design. Modern military systems (and information technology systems) are so complex that only specialists can understand the detailed operations of the protocols, identify critical vulnerabilities, and understand how to address these vulnerabilities without compromising functionality. Many of these details are specific to each military system, and therefore the technical knowledge is confined to a very limited number of experts.

Second, functionality and cybersecurity are intertwined. Quite a number of cyber vulnerabilities are the result of features deliberately designed into systems. That is not to say that engineers aim to make vulnerable systems, but during design, engineers make trades between functionality and security and are willing to accept certain levels of vulnerabilities in order to achieve some functionality, often knowingly, and sometimes unknowingly. One example is that desktop computer operating systems permit other computers to download and run executable code from World Wide Web sites. Much of the functionality of the World Wide Web comes from this ability, which allows for a powerful range of applications. But it also permits the loading of malware from malevolent actors. Much of the commercial world is so driven by introducing new functionality that security is a lesser priority, and, when addressed, security is introduced by overlays on an insecure design rather than by an inherently secure design.

---

[7] See also Baldwin, Dahmann, and Goodnight, 2011.

Further, because the most common means of gaining access to cyber systems is to exploit a combination of a design feature (e.g., enabling the running of an external executable code) and poor human behavior (e.g., opening an email attachment from a suspect source), cybersecurity measures also tend to take the form of restrictions on the operator and hence interfere to varying degrees with the operation of the system and the accomplishment of the mission.[8] Operators see clearly how these restrictions make their operational mission more difficult; they often do not see the dangers the restrictions are intended to mitigate. By defining concepts of operations that include how operators interact with systems, operational commanders, then, also make decisions that affect the balance of operational functionality and security, further intertwining functionality and cybersecurity.

Third, the threats of exploitation and attack through cyberspace are rapidly evolving and adapting to countermeasures. Capabilities of potential adversaries are growing, and the changing technologies introduce new vulnerabilities over time. This evolution means that static solutions for cybersecurity management are unlikely to be effective; cybersecurity solutions need to be adaptive. Creating defensive barriers in the form of security overlays that respond to discovered vulnerabilities is by nature insufficient to protect against future, unknown threat vectors.

Fourth, the offense (in this case, U.S. adversaries) has some advantages over the defense (in this case, the United States). The adversary needs to find only one weakness, such as a single access point. The defender needs to mitigate against all plausible threats for all potential vulnerabilities. Because the aggressor needs to find just one path for access and the defender needs to block all possible paths, defensive measures also tend to be more expensive than the tools they attempt to stop, putting the defensive side on the wrong side of cost considerations.

Fifth, this imbalance between the offense and defense in the cyber domain implies that it is unwise to assume that complete cybersecurity can be achieved.[9] Some potential vulnerabilities that can be exploited or attacked will always persist. The goals of counter cyber exploitation are, for example, controlling critical information by identifying it, restricting access to it, and preventing its theft. It is not possible to reduce the amount of critical information to zero. Nor does it appear safe to assume that access can be unequivocally denied. The question is how much security is enough given finite resources and mission needs.

Sixth and finally, risk mitigation decisions for cybersecurity are not easily partitioned. Systems are sufficiently interconnected that accepting (either knowingly or unknowingly) a vulnerability in one system can introduce a vulnerability into another system. For example, accepting risk of an adversary accessing a maintenance device through an intermittent Internet connection potentially introduces this vulnerability to a military system when that maintenance device is connected to the military system. In a highly interconnected system of systems, this

---

[8] See, for example, National Institute of Standards and Technology (NIST) Special Publication 800-53, April 2013.

[9] For some recent overviews of the cyber threat that make this point, see Krekel, Adams, and Bakos, 2012; Maybury, 2012; and Gosler and Von Thaer, 2013.

Achilles' heel can be remote from the most critical component or most critical system. An aircraft could be accessed, for example, by targeting its support equipment. Access for information about a system can be gained outside the Air Force, for example, by targeting the industrial base. Hence, critical vulnerabilities can be introduced in small programs or in noncritical components and in a diverse range of targets. Concentrating effort on vulnerability reduction in only large or critical systems is insufficient.

General Management Challenges

In addition to these challenges that inhere in cybersecurity, the management of cybersecurity faces the same challenges as any other management function. These challenges extend to all dimensions of DOTMLPF-P, but for emphasis, we specifically highlight two of these challenges.

First, cybersecurity, like any other aspect of life-cycle management, needs to be covered in all the (temporally overlapping) phases of the military system life cycle. In the research and development phase, decisions shape architecture and other structural matters that can render a system design easy or difficult to secure and might result in cost savings relative to trying to impose security on a system after development. The procurement phase sets derived requirements and places them on contract to industry. Developmental and operational testing validate that requirements have been met and assess whether systems are operationally effective, suitable, survivable, and safe for intended use. During the long sustainment phase, changes to the military system, concepts of operation, and the threat environment require continuous vigilance in cybersecurity. And finally, disposal must be done in a way that does not compromise other systems, for example, by revealing vulnerabilities. Cybersecurity management needs to cover this cradle-to-grave life cycle.

The goals outlined in the last section of what needs to be done to achieve an acceptable level of cybersecurity are most easily and effectively done when integrated with the design phase. Poor system security engineering is very difficult to mitigate by overlaying security controls, whereas security controls overlaid on a sound, secure design can be quite effective.[10] For systems that are fielded and no longer in production, design changes to improve cybersecurity generally necessitate a modification program and can be cost-prohibitive. Most Air Force systems reside in this "legacy" phase. It is especially important in this phase that a mission assurance perspective be adopted that examines the full spectrum of options for cybersecurity, including after-design protective measures, changes in operational procedures, and modifications, if necessary and affordable. Those modifications could include design changes in ware and changes in how humans behave when operating the systems.

---

[10] See, for example, Anderson, 2008; Bayuk, Healy, et al., 2012.

Second, monitoring and feedback are necessary for the management of any enterprise.[11] They are most useful when tailored to the decisions each level in an organization needs to make. Those who make decisions regarding cybersecurity need feedback on the performance of military systems and missions under their control to make adjustments and inform better decisions in the future. *Feedback* is the process that delivers an evaluation of the state of cybersecurity for the purposes of learning and control. *Metrics* are quantitative measures that express cybersecurity performance so that the deviation from the desired performance can be assessed, thus providing a cue for the adjustment of cybersecurity designs and practices.

In the next section, we discuss the implications of these challenges for managing the cybersecurity of military systems throughout their life cycles.

Principles for Managing Cybersecurity

General Considerations

Three of the cybersecurity challenges discussed above—(1) the technical nature of cybersecurity solutions, (2) the degree to which cybersecurity and operational functionality are intertwined in design, and (3) the involvement of human factors in cybersecurity—point to the need to assign these decisions to engineers. Staff and higher-level officials do not possess the expertise to make effective decisions in this area and are not intimately involved in the systems engineering process when these decisions need to be made. Mitigating cyber exploitation, then, entails integrated system design for more secure architectures and consideration of how human operators are likely to interact with systems in the field.[12] The cybersecurity solutions for military systems also necessarily need to be tailored on a case-by-case basis.

The kind of engineering that addresses such security concerns has been called *system security engineering*, defined by the DoD in 1995 as

> an element of system engineering that applies scientific and engineering
> principles to identify security vulnerabilities and minimize or contain risks
> associated with these vulnerabilities. It uses mathematical, physical, and related
> scientific disciplines, and the principles and methods of engineering design and
> analysis to specify, predict, and evaluate the vulnerability of the system to
> security threats.[13]

---

[11] We will use the term *monitoring* to mean the effective collection of information on the performance of an enterprise. *Feedback* is the process of disseminating this information in a form that leaders can use for decisionmaking.

[12] Bayuk and Horowitz, 2011; Reason, 1990.

[13] Department of Defense Handbook MIL-HDBK-1785, 1995, p. 5.

System security engineering is a critical component in achieving effective cybersecurity.[14] It is outside the scope of this project to describe the tradecraft of system security engineering; we refer the reader to the recent literature on the subject.[15] Our focus is how to arrange roles and responsibilities, incentives, authorities, monitoring, and accountability in the Air Force to improve cybersecurity overall, including how to best encourage and support system security engineering in the Air Force.

Two of the cybersecurity challenges discussed above—(1) the imbalance between offense and defense in the cyber domain and (2) that risk mitigation is not easily partitioned—indicate that no perfect solution is likely to be found and that multiple communities have a stake in managing this risk assessment. These communities include, at minimum, the operational user of the military system (who has visibility into concepts of operations and how the military system combines with others to carry out a mission, and suffers the consequences if it does not), the intelligence community (which assesses the threats), and the program office (which oversees the design and testing, sets requirements for system security engineering, and has overall responsibilities for life-cycle management). The better coordinated and integrated these efforts, the better the opportunities for effective cybersecurity.

The cybersecurity challenges of the evolving threat environment and the changing technologies stress the need for adaptive solutions and an organization that supports and facilitates innovation. These challenges contraindicate that effective cybersecurity solutions will emerge solely from prescribed rules and policies, specifically in the form of security controls. Reactive, defensive barriers meant to deny access have struggled to keep up with the evolving threat.[16] Although the most common means to try to reduce cyber exploitation is defensive barriers to deny access, these barriers are most effective when they are part of the integrated design of the system than when they are appended after design.[17] Security controls enveloping a system poorly designed from a security standpoint are unlikely to be successful; controls integrated with a secure architectural design have great promise. Effective cybersecurity management is most likely to be achieved through risk mitigation guided by mission assurance goals and achieved by adaptive solutions integrated into the design phase, rather than by prescribed rules.

---

[14] Some aspects of human factor engineering play a role in cybersecurity. For example, how humans really operate a system can differ from its expected operation in ways that make it less secure. Real people, for example, forget complex, frequently changing passwords and sometimes write them down where they are not secure, or those passwords often get changed when forgotten to a temporary administrative default that is not as secure. Taking these human behaviors into account in design is included in our use of the term *system security engineering*.

[15] See, for example, Anderson, 2008; Bayuk, Barnabe, et al., 2010; and Ross, Oren, and McEvilley, 2014.

[16] Gosler and Von Thaer, 2013.

[17] Bayuk and Horowitz, 2011.

Organizational Roles and Responsibilities

No ideal organizational design exists.[18] Any organizational structure and assignment of roles and responsibilities facilitate certain outcomes and impede others. Enterprises organize accordingly, and there are numerous models of organizations that attempt to rationalize organizational design and outcomes.[19] Hence, the literature on organizations does not point to any unequivocal solutions for how cybersecurity objectives should be organized. Nevertheless, certain characteristics of organizational design are accepted as good practice, and we will draw on these to shape how organizational roles and responsibilities might be assigned to facilitate cybersecurity.

The organizational sociologist Henry Mintzberg describes three characteristics (relevant to cybersecurity management) of the environment in which an enterprise operates that affect how it organizes:[20]

1. How stable the environment is. If the problems that the enterprise faces do not change much over time, and the changes it faces are predictable, the environment is said to be *stable*. The opposite, an environment that changes unpredictably, Mintzberg calls *dynamic*.
2. How complex the environment is. If the problems faced by the enterprise are challenging and require specialized knowledge for finding their solutions, the environment is said to be *complex*. The opposite, an environment of relatively easy problems that do not require deep knowledge, Mintzberg calls *simple*.
3. How diverse the environment is. If the enterprise faces a variety of problems and conducts a wide range of activities to meet these needs, the environment is said to be *diverse*. The opposite, an environment that presents few problems and requires few distinct activities, Mintzberg calls *integrated*.

Where the environment lies in the spectrum between the extremes of these three characteristics is highly correlated with how enterprises organize to cope with these environments. These tendencies are shown graphically in Figure 1.1. The more simple its environment is, the more an organization can predict the challenges faced by its members, and the more those challenges are stable. Therefore, the organization can operate efficiently by standardizing and formalizing in policy how its members should perform their tasks. The more dynamic the environment is, the less the enterprise can predict the nature of the exact tasks, and the more rapidly those tasks change. The tendency in dynamic environments is for the enterprise

---

[18] Following Snyder et al., 2013, we use the term *organizational design* to mean "a set of parameters that determine the division of labor and specify how coordination should occur within and across those divisions to promote efficient, effective flow of work, information, and decisions throughout the organization."

[19] See, for example, Scott, 1992.

[20] Mintzberg, 1979, pp. 267–270.

to organize in a way that moves from standardization of tasks as a means of control to one that coordinates task execution via collaborative interactions with others.[21]

Figure 1.1. How Environment Affects How Enterprises Organize

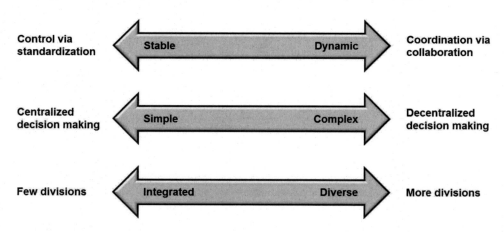

The more complex the environment, the greater and deeper the knowledge needed to perform tasks. As this knowledge is located throughout the organization, decisions tend to be decentralized to the locus of knowledge appropriate for task execution, and the organization manages task execution via coordination. When the environment is simple, organizations tend to centralize control at high echelons in the organization in order to gain efficiencies in operations. In the extreme case of simplicity, if a single person can understand the full nature of the operation, it is more efficient and effective to have that one person control how tasks are performed, thus alleviating the need for communication and coordination.[22]

When the environment is diverse, the tasks performed vary considerably, and the organization tends to split into divisions organized around common tasks. For example, if consumer tastes and regulations for an automobile are considerably different in the United States, Europe, and Japan, a company that has markets in these regions can be more effective by creating divisions around these markets in order to focus effort and be responsive to the specific needs of these markets. Organizations facing integrated environments tend not to cleave into such divisions in order to retain economies of scale. For organizing around common tasks to be effective, the benefits of splitting into divisions must outweigh the loss of economies of scale.[23]

We argue that the fast-changing technologies and the evolving and adapting threat environment place cybersecurity toward the dynamic end of the continuum of the first characteristic, stability. We further argue that all six of the management challenges peculiar to

---

[21] Mintzberg, 1979, pp. 270–272, and literature cited therein. The tendency to coordinate task execution in dynamic environments is stronger than the tendency to standardize task execution in stable environments.

[22] Mintzberg, 1979, pp. 273–277, and literature cited therein; Damanpour, 1991; Mihm et al., 2010.

[23] Mintzberg, 1979, pp. 278–281, and literature cited therein.

cybersecurity render it complex. And finally, given the diversity of military systems, information systems, and the peculiar security issues they face, we place cybersecurity a little more toward the diverse than the integrated side of the continuum.

The tendency toward a dynamic environment suggests that cybersecurity is most effectively managed by an organizational design that emphasizes identifying cybersecurity solutions via coordination and collaboration of workers rather than prescribing standardized solutions. The strong tendency toward complexity suggests that it is unlikely that the organization will identify successful solutions from the top down. Highly formalized rules for achieving cybersecurity imposed on the engineering level from above will impede innovative, adaptive outcomes for cybersecurity. Decentralized decisionmaking for implementation to the levels in the organization possessing the appropriate expertise is indicated. Higher levels in the hierarchy should set clear goals of what should be accomplished, but devolve decisions of how to accomplish those goals to well-trained system security engineers. And finally, the weaker tendency toward diversity suggests that cybersecurity might very well require more than one organizational unit to be effectively managed, perhaps managing weapon systems/platform information systems and information technology systems separately.

Feedback and Control

Continuous monitoring of how well desired outcomes are being achieved is essential for sound decisionmaking. What is monitored and evaluated in an organization sets incentives and accountability structures and therefore can implicitly set the organization's goals.[24] If those implicit goals do not align with the organization's desired outcomes, it is unlikely those outcomes will be attained.

Cybersecurity management can be understood as a process-control loop.[25] As depicted by the solid arrows in Figure 1.2, a process-control loop runs from the state of cybersecurity, to feedback mechanisms for monitoring it, to decisionmakers, to the actions (designs, policies, and practices) under their control, which in turn adjust the state of cybersecurity. This process-control loop depicts standard management practices of a continuous process of monitoring how well the enterprise is operating, making decisions to adjust those operations, taking actions to adjust the enterprise, then examining the outcomes of those decisions. There is also a feedback loop that monitors whether the actions directed by management are being properly implemented. This compliance feedback is depicted in Figure 1.2 by the dashed arrow.

---

[24] Jensen and Meckling, 1992.

[25] This is a commonly used framework for management decisions dating back at least to the work of Steinbruner, 1974.

Figure 1.2. Depiction of a Process-Control Loop for Managing Cybersecurity

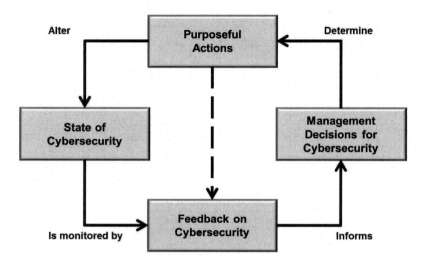

If an enterprise monitors only compliance with its directives (the dashed arrow) and is not effective in monitoring the real state of the enterprise (the solid arrow), it cannot effectively control that state, and the organization sends a message to its members that compliance is the enterprise objective, not performance outcomes. The effective monitoring and feedback of the real state of an enterprise requires two basic elements: (1) a baseline, defined by outcome-based objectives for the enterprise, from which to evaluate the state and (2) mechanisms that comprehensively monitor and provide feedback on the real state of the enterprise. Those mechanisms must provide feedback that informs decisions that each level in management is empowered to make.

Decisionmakers at high levels in the Air Force need feedback on cybersecurity that captures the success of cybersecurity measures in meeting mission assurance goals and need to look out at a longer timeframe, monitoring trends over time. Decisionmakers at lower (engineering) levels need more technical feedback regarding the performance of systems and current threats, and need more rapid responses than the higher-level leaders do.

The goals and priorities for cybersecurity and mission assurance are important for shaping the right feedback for highest-level leaders. In any endeavor, it is easy to take one's eye off the ultimate goal in favor of more proximate objectives. A common pitfall of analysis is to adopt as goals the most clearly defined and easily measured phenomena. James Roche and Barry Watts note a common military example of this mistake—using attrition rate as a measure of success in battle. Attrition rate is generally a poor metric, because victory in a given battle is only weakly correlated with attrition. More commonly, some other factor decides the outcome, such as outmaneuvering the adversary. As they note, "In military affairs, the most obvious or readily quantifiable measures may not necessarily be the right ones at all."[26] For this reason, it is

---

[26] Roche and Watts, 1991.

14

important to clearly articulate the desired outcomes of cybersecurity and let those goals direct the selection of feedback and metrics, what we call *outcome-based feedback*, rather than letting the attributes that are most easily measured implicitly dictate the goals.

It is also vital to understand the limitations of feedback and how it is best used in decisionmaking, especially when attempting to quantify feedback in the form of metrics. Like safety, the nature of firm feedback in cybersecurity is mostly negative—that cybersecurity measures have failed. Negative feedback is clear—a problem exists that needs remedy. Positive feedback regarding the state of the enterprise, taking the form of not finding clear cybersecurity failures, might or might not mean that the system is adequately secure. In the case of red teams, exercises, and operational testing, it might just mean that the teams did not apply the same resources and comparable skills into trying to defeat the system as an adversary might; in the case of intelligence, it might mean that the adversary has already been successful, but that it has not been detected. Positive feedback regarding compliance gives incomplete insight into the actual state of the enterprise unless the rules are guaranteed to prevent the problems.

Hence, negative and positive feedback need to be assessed and used differently, as is done in the field of safety. The negative feedback can be used for learning and correction.[27] Safety investigations and reports are produced only to foster learning and prevent future mishaps, so much so that they take precedence over accident investigation boards.[28] Positive feedback, whether from good compliance or lack of evidence of real cybersecurity issues, can lull leaders into unwarranted complacency. This tendency for complacency might be more acute for cybersecurity than safety because the positive feedback might be due more to ignorance than reality. For safety, complacency is battled by instilling safety as a core value in an organization, resulting in constant vigilance and attention to safety. If feedback on safety were restricted to compliance with checklists, leaders would tend to be lulled into complacency.

For these reasons, feedback in cybersecurity can benefit a lot from supplementing passive monitoring with active probing in the form of operational tests, red teams, and cybersecurity exercises. These can scrutinize the systems and how they are used operationally to reveal vulnerabilities and promote enterprise learning. But military systems and cybersecurity are very complex, so the direct causes and effects of failed cybersecurity measures will in general be challenging to establish for complex systems. These attributes place additional limitations on the feedback, and hence place a high burden of expertise and tacit knowledge of cybersecurity on decisionmakers. Expertise in the systems and details of cybersecurity are important for the

---

[27] But if orders or laws are violated, actors are still held accountable. Learning and accountability are administered separately.

[28] Air Force Instruction 91-204, 2014.

proper interpretation of feedback. It also emphasizes the need for qualitative feedback to supplement quantitative metrics.[29]

Finally, outcome-based feedback for cybersecurity must express the impact on the operational mission. Compliance-based metrics and feedback fail to do this on two accounts. First, they monitor how well the decisions are being implemented (the upper box in Figure 1.2) and not the state of cybersecurity. Second, they monitor system-level vulnerabilities and not mission-level impact. For cybersecurity to have the appropriate level of priority for leaders, and to hold them accountable, feedback needs to capture the current state of cybersecurity and how that affects operational missions.

Summary

To summarize, cybersecurity policy needs to go beyond *how* to do cybersecurity by specifying *what* needs to occur through clear, outcome-based objectives. We propose that the outcome that is desired is to limit adversary intelligence exploitation through cyberspace to an acceptable level and to ensure an acceptable level of operational functionality (survivability) even when attacked offensively through cyberspace. What constitutes an acceptable level of risk would be guided by mission assurance risk acceptance. These outcomes need to be achieved continuously throughout the life cycle of a military system, all the way from research and development through disposal. All phases are important, but the development stage is critical because design decisions are made that can limit options in the future, and the sustainment phase is critical because most systems reside in that phase and remain there for extended periods.

Cybersecurity risk management has three components: (1) minimizing vulnerabilities to systems, (2) understanding the threat to those systems, and (3) minimizing the impact to operational missions. Multiple layers contribute to mitigating vulnerabilities: defensive measures to deny access to the systems, backed up by a robust and resilient design, so that, when attacked, the system degrades gracefully and recovers rapidly to an acceptable level of functional performance. Robustness and resiliency are inherent in the design of a system. The overall operational risk reduction will come from a combination of system security engineering, assessment of how mission assurance is affected, and, because the cybersecurity environment is rapidly changing, adaptive solutions. Finding such solutions will require the integration and coordination of efforts across multiple stakeholders, with mission assurance as the central guide.

The cybersecurity environment is dynamic and complex, and organizations cope with such environments by choosing organizational designs that identify solutions via coordination and collaboration of workers rather than prescribing standardized and formalized solutions dictated from the top down. They favor instead decentralized processes for identifying solutions to drive

---

[29] Qualitative feedback could take a form similar to the feedback provided in commercial aviation through the Aviation Safety Reporting System; see Reynard et al., 1986; and Vicente, 2003, pp. 195–203.

innovation and adaptive outcomes. To manage cybersecurity and hold actors accountable, outcome-based feedback is needed to measure the state of cybersecurity and its mission impact, not just compliance with directives.

These observations form the baseline against which we evaluate the current state of cybersecurity management in the Air Force. The following chapter will describe the current state of cybersecurity policy and its implementation in the Air Force from this perspective.

2. Cybersecurity Laws and Policies

"The errors of a theory are rarely found in what it asserts explicitly; they hide in what it ignores or tacitly assumes."

– Daniel Kahneman[1]

Introduction

The governance structure for cybersecurity defined by the laws and policies at the federal, DoD, and Air Force levels is voluminous and complicated. Even summarizing this body of oversight would require a document of considerable length. We do not attempt a comprehensive review, but instead describe this governance structure in light of the principal issues raised in Chapter One. We first highlight key facets of legislation and federal and DoD policy, including some relevant history, to identify the stakeholders, requirements, and constraints within which Air Force cybersecurity must operate. The second part of the chapter outlines cybersecurity practice within the Air Force, paying particular attention to roles and responsibilities and feedback.

One important aspect of the laws and policies is that they change frequently relative to the duration of the life cycle of a military system. In practice, this frequent reworking of the governance structure for cybersecurity means that military systems often fall under multiple governance structures over the course of their life cycles. As we document in this chapter, cybersecurity governance has changed significantly every several years, whereas many weapon systems have service lives in the decades. Indeed, some systems predate most of the modern governance structures and, while in sustainment, have to varying degrees been managed under multiple cybersecurity frameworks. These temporal changes therefore present a challenge to assessing the success or shortcomings of laws and policies, because many of the laws and policies have not been in effect long enough to correlate with specific outcomes.

Another aspect of the evolving governance structure is a non-uniformity in the use of terms within DoD over time, especially the terms *information assurance* and *cybersecurity*. In general, legislation relevant to the subject of this report refers to *information security* rather than the more recently introduced term *cybersecurity*. As of 2014, DoD policies have shifted to using the term *cybersecurity*. Some policies replace *information security/assurance* with *cybersecurity* without actually changing meaning. Many residual policies remain that use the term *information assurance*, defined by DoD as "actions that protect and defend information systems by ensuring availability, integrity, authentication, confidentiality, and nonrepudiation."[2] In this chapter, when

---

[1] Kahneman, 2011, pp. 274–275.

[2] Joint Publication 1-02, 2014. For the definition of *information security* used in legislation, see 44 U.S.C. § 3532.

discussing the governance structure, we generally use the term *cybersecurity* except in quotations and certain historical discussions. The various terms as used in law and policy will be treated as identical in meaning except where explicitly stated otherwise.

Legislation and Federal Cybersecurity Policy

Through legislation, Congress shapes the government's cybersecurity efforts. This body of law assigns some specific roles and responsibilities within the government, creates offices and reporting structures, and specifies a number of practices that agencies must follow regarding cybersecurity. This section outlines some of these prescriptions.

The primary pieces of legislation regarding cybersecurity are the Paperwork Reduction Act of 1980,[3] the Clinger-Cohen Act of 1996,[4] and the Federal Information Security Management Act of 2002 (FISMA), Title III of the E-Government Act of 2002.[5] The Paperwork Reduction Act charged the Office of Management and Budget with oversight responsibilities for federal agency information resource management, which includes acquisition and use of information technology. Under the Paperwork Reduction Act, the Office of Management and Budget was given responsibility for developing information security policies. The Clinger-Cohen Act, which comprised the Information Technology Management Reform Act and the Federal Acquisition Reform Act, established the position of chief information officer (CIO) in government agencies, including DoD and the military departments, and assigned the CIO position a set of responsibilities over information technology management and information security. Those responsibilities for DoD and Air Force CIOs will be described later in this chapter.

FISMA directs each government agency to implement an information security program, based on periodic risk assessments. It requires agency-wide programs to "cost-effectively reduce information security risks to an acceptable level" and "ensure that information security is addressed throughout the life cycle of each agency information system."[6] Note that this last provision addresses information systems directly, rather than agency-wide or mission-level security. FISMA further directs NIST to "develop standards and guidelines, including minimum requirements, for providing adequate information security for all agency operations and assets."[7] This charge includes developing risk-based categorization standards for information and information systems and "minimum information security requirements" for each category.[8] The

---

[3] Pub. L. 96-511, 1980.

[4] Pub. L. 104-106, Divisions D and E, 1996; and 40 U.S.C. § 11101 et seq.

[5] Pub. L. 107-347, 2002. For a comprehensive listing of legislation, see Fischer, 2013.

[6] 44 U.S.C. § 3544.

[7] 15 U.S.C. § 278g-3. Similar provisions were included in the Clinger-Cohen Act; FISMA expands and adds detail to NIST's information security mandate.

[8] 15 U.S.C. § 278g-3.

standards and guidelines are required by the act "to the maximum extent practicable" to permit the use of commercial, off-the-shelf information security software and to "provide for sufficient flexibility to permit alternative solutions" to address information security risks.[9] In response to FISMA, NIST developed two Federal Information Processing Standard documents supported by a number of Special Publications that define the "Standard for Minimum Security Requirements for Federal Information and Information Systems."[10] Later, NIST expanded their guidance by creating an encompassing Risk Management Framework, the nature and implementation of which are defined in a set of Special Publications and will be discussed in more detail later in this chapter.[11] FISMA gives the Office of Management and Budget oversight over all agency "information security policies and practices."[12] However, these responsibilities for mission-critical systems in DoD are delegated to the Secretary of Defense.

National Security Systems, Platform Information Technology (PIT), and PIT Systems

The legislation discussed above focused on improving acquisition and management of administrative and business information technology systems and networks, and carved out certain exemptions for *national security systems* (NSS), which are defined in the Clinger-Cohen Act as follows:

> a) Definition.—
>
> (1) National security system.— In this section, the term "national security system" means a telecommunications or information system operated by the Federal Government, the function, operation, or use of which—
>
> (A) involves intelligence activities;
>
> (B) involves cryptologic activities related to national security;
>
> (C) involves command and control of military forces;
>
> (D) involves equipment that is an integral part of a weapon or weapons system; or
>
> (E) subject to paragraph (2), is critical to the direct fulfillment of military or intelligence missions.

---

[9] 15 U.S.C. § 278g-3.

[10] The primary documents are NIST, *Standards for Security Categorization of Federal Information and Information Systems*, Federal Information Processing Standard Publication 199, February 2004; NIST, *Minimum Security Requirements for Federal Information and Information Systems*, Federal Information Processing Standard Publication 200,March 2006; and NIST Special Publication 800-53, *Security and Privacy Controls for Federal Information Systems and Organizations*, 2013.

[11] The master documents are NIST Special Publication 800-39, *Managing Information Security Risk: Organization, Mission, and Information System View*, March 2011; and NIST Special Publication 800-37, *Guide for Applying the Risk Management Framework to Federal Information Systems: A Security Life Cycle Approach*, Revision 1, 2014.

[12] 44 U.S.C. § 3543.

(2) Limitation.— Paragraph (1)(E) does not include a system to be used for routine administrative and business applications (including payroll, finance, logistics, and personnel management applications).[13]

The Clinger-Cohen Act exempts NSS from most of its provisions, but the agency CIO responsibilities and authorities defined in the act do apply to NSS. The NIST information security standards and guidelines described in FISMA do not apply to NSS, freeing agencies to create separate information security policies for general information systems and for NSS.[14]

These statutory exceptions implicitly acknowledge that NSS generally differs from business and administrative information technology systems and networks and that agencies may choose to manage NSS cybersecurity separately. However, this freedom is limited both by other FISMA provisions and by current national policy. FISMA directs the Office of Management and Budget to coordinate the development of standards and guidelines between NIST and agencies controlling NSS such that they are as complementary as possible.[15] The Committee on National Security Systems (CNSS), which is chaired by the DoD CIO, requires in policy that all organizations with NSS implement "an [information assurance] risk management program for their NSS that . . . is consistent with the provisions of NIST SP 800-39 [*Managing Information Security Risk: Organization, Mission, and Information System View*]," the foundational document of the NIST information security methodology.[16]

DoD policies on cybersecurity in general do not reference NSS. They do, however, define a category called *platform information technology* (PIT), which has significant overlap with NSS. PIT is defined as "[information technology], both hardware and software, that is physically part of, dedicated to, or essential in real time to the mission performance of special purpose systems."[17] Further, a *PIT system* is defined as "a collection of PIT within an identified boundary under the control of a single authority or security policy."[18] The distinction between PIT and PIT system is significant: Whereas a PIT system requires cybersecurity authorization in order to operate, PIT does not. The distinction between PIT and PIT system is determined on a case-by-case basis.[19] Although exceptions exist, for the purposes of this report, we assume that all information technology associated with military systems qualify both as NSS and PIT (or a PIT system) and therefore are subject to the laws, policies, and guidance relevant to those categories.

---

[13] 44 U.S.C. § 3502.

[14] 15 U.S.C. § 278g-3.

[15] 44 U.S.C. § 3543.

[16] CNSS Policy 22, 2012, p. 3.

[17] DoD Instruction 8500.01, 2014, p. 58.

[18] DoD Instruction 8500.01, 2014, p. 58.

[19] DoD Instruction 8500.01, 2014, p.40.

Chief Information Officers

Congress mandates that agencies have a CIO whose responsibilities are defined by the Clinger-Cohen Act and FISMA. In general, the agency CIO is broadly responsible for providing advice and assistance to the agency head for information technology acquisition and information resource management, as well as ensuring that the agency has a sound and integrated information technology architecture.[20] Although agency heads are ultimately responsible for information security within their agencies, they are required by FISMA to delegate to the CIO the authority to ensure compliance. The CIO is thus responsible for developing and maintaining the agency's information security program, including NSS and other information systems. The CIO is also authorized to appoint a senior agency information security officer who carries out these information security–related responsibilities.[21]

NIST Risk Management Framework

High-level governance goes beyond defining desired outcomes for cybersecurity by prescribing how agencies must achieve cybersecurity. The Clinger-Cohen Act,[22] FISMA,[23] CNSS policy,[24] and current DoD policy[25] mandate that the cybersecurity programs implemented by government agencies adhere to (or, in the case of NSS, are "consistent with") NIST-developed standards and guidelines, specifically with NIST SP 800-39, *Managing Information Security Risk: Organization, Mission, and Information System View*. This document outlines a risk management approach that evaluates risks at three tiers. Tier 1 evaluates risks at the organizational level by establishing governance structures, defining priorities, developing investment strategies, and establishing common security controls for the enterprise. Tier 1 is the responsibility of the agency head and CIO. Tier 2 evaluates risk at the mission/business level by defining processes and prioritizing those processes according to the guidance from Tier 1. Tier 2 is the responsibility of mission and business owners. Tier 3 evaluates risk from an information system perspective.[26] Most of the detailed policy for cybersecurity risk management treats the system level in Tier 3 via the NIST Risk Management Framework (RMF).[27] This overall risk

---

[20] 40 U.S.C. § 11315.

[21] 44 U.S.C. § 3544.

[22] 40 U.S.C. § 110302.

[23] 44 U.S.C. § 3543.

[24] CNSS Instruction 1253, 2012, pp. 2, 8.

[25] DoD Instruction 8510.01, 2014, p. 2.

[26] NIST Special Publication 800-39, 2011, pp. 9–11.

[27] NIST Special Publication 800-37, 2014.

management methodology is consistent with DoD's Mission Assurance Strategy.[28] We discuss the NIST methodology in more detail later in this chapter, with particular focus on Tier 3.

Department of Defense Cybersecurity Policy

DoD cybersecurity policy has a long and evolving history. It is helpful to understand some of this history for two reasons. First, early versions of the policy have to a large extent defined the framework within which current policy was developed and is implemented. Second, most DoD military systems are in sustainment and were therefore acquired under earlier policies (or in the case of some old systems, before there was a formal policy). The summary in this section is necessarily brief and is therefore not comprehensive.

Before 2007

Information security within DoD has a long history prior to the Paperwork Reduction Act. As early as 1972, DoD issued policy on "Security Requirements for Automated Data Processing Systems,"[29] which was replaced in 1988 by "Security Requirements for Automated Information Systems."[30] In accordance with these policies, DoD developed the DoD Information Technology Security Certification and Accreditation Process (DITSCAP).[31] The DITSCAP applied to information technology system acquisition and life-cycle management; organization- and mission-level cybersecurity risk considerations were not explicitly included.

DITSCAP centered around a defined set of security requirements: accountability, access, security training and awareness, physical controls, marking, least privilege, data continuity, data integrity, contingency planning, accreditation, and risk management. DITSCAP identified four key roles for authorizing information assurance risk acceptance (roles that have not changed appreciably since):[32]

- The *designated approving (or accrediting) authority* (DAA) was a senior official assigned responsibility by the component head for ensuring systems operate at an acceptable level of residual information assurance risk.
- The *certification authority* (CA) was the subject matter expert responsible to the DAA who assesses the level of security compliance and risk associated with a system and makes an authorization recommendation to the authorizing official.
- The program manager is responsible for ensuring that the required security controls and safeguards are implemented to satisfy the requirements and obtain authorization. The program manager is ultimately responsible for the life-cycle management of the system.

---

[28] U.S. Department of Defense, 2012.

[29] DoD Directive 5200.28, 1972.

[30] DoD Directive 5200.28, 1988.

[31] DoD Instruction 5200.40, 1997.

[32] DoD Instruction 5200.40, 1997, p. 14.

- The *information technology system user representative* is the individual or organization that represents the user or user community in the definition of information system requirements.

DITSCAP was the DoD framework for authorizing information assurance risk acceptance from 1997 to 2007.

2007–2014

DITSCAP was replaced in 2007 by the DoD Information Assurance Certification and Accreditation Process (DIACAP).[33] The DIACAP retained a focus on individual systems, but instituted a number of changes from DITSCAP, including the following:

- Under DITSCAP, categories of security requirements were identified, but the specific requirements themselves were not defined. In DIACAP, the term *security requirements* was generally replaced with *information assurance controls* and, as required in FISMA, an enumerated set of NIST-defined controls was adopted.[34] With this change, security requirements were clearly defined as that which are satisfied by the application of a subset of the list of information assurance controls. Information assurance controls, now called *security controls*, will be discussed further in the next section.
- PIT was brought into the DoD information assurance system. At this juncture, only "PIT interconnections," or network access to PIT, required authorization. In general, PIT that was not directly connected to communication networks was considered to be exempt from DIACAP certification and authorization.[35]

DIACAP was the DoD framework for authorizing information assurance risk acceptance from 2007 to 2014.

2014 and Beyond

In March 2014, DoD shifted emphasis from the previous goal of information assurance in the DIACAP policy to a goal of cybersecurity in the DoD RMF.[36] In the DoD RMF, the new positions of authorizing officials and security control assessors take on parallel roles of the previous DAAs and CAs. For the first time, the concept of multi-tiered risk management, which includes assessment of risk at organization and mission levels in addition to the system level, was introduced into the DoD cybersecurity program. The new policy assigns governance responsibilities for Tier 1 and Tier 2 risk management, but not detailed procedures for how these responsibilities should be accomplished. The DoD CIO is ultimately responsible for Tier 1 risk management, and the component CIOs are responsible for Tier 2.[37] Implementation of

---

[33] DoD Instruction 8510.01, 2007, reissued and renamed March 12, 2014.

[34] DoD Directive 8500.01E, 2007.

[35] DoD Directive 8500.01E, 2007, p. 2.

[36] Defined by DoD Instruction 8500.01, 2014; and DoD Instruction 8510.01, 2014.

[37] DoD Instruction 8510.01, 2014, p. 14.

cybersecurity is still largely restricted to the system level, however: "Though the need for specific protections is identified at Tiers 1 and 2, it is at Tier 3 where the information protections are applied to the system and its environment of operation for the benefit of successfully enabling mission and business success."[38] The implementation of Tier 3 risk management follows the NIST RMF.[39] Note that the DoD RMF adopts and encompasses all three tiers of the NIST risk management methodology, while the NIST RMF focuses primarily on Tier 3.

Also for the first time, the new DoD RMF explicitly requires cybersecurity to address PIT and PIT systems. Although individual PIT does not necessarily require an authorization to operate, "a collection of PIT" that the system owner and authorizing official agree "rises to the level of a PIT system" does, as outlined in Figure 2.1.[40]

Figure 2.1. DoD Information Technology

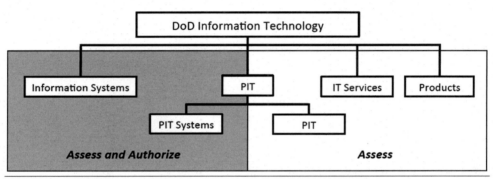

SOURCE: Modified from DoD Instruction 8500.01, March 14, 2014.

The NIST RMF process is built around identifying and implementing appropriate "security controls," which are "safeguards or countermeasures prescribed for an information system to protect the confidentiality, integrity, and availability of the system and its information."[41] Further, "information security requirements are satisfied by the selection of appropriate . . . security controls from NIST SP 800-53 [*Security and Privacy Controls for Federal Information Systems and Organizations*]."[42] Policy does state that "these controls include ensuring that developers employ sound systems security engineering principles and processes including, for example, providing a comprehensive security architecture."[43] However, the first step of the NIST RMF process is "categorize the information system,"[44] implying that the system is already

---

[38] DoD Instruction 8500.01, 2014, p. 29.

[39] DoD Instruction 8510.01, 2014, p. 2.

[40] DoD Instruction 8500.01, 2014, p.8.

[41] NIST Special Publication 800-37, 2014, p. B-9.

[42] NIST Special Publication 800-37, 2014, p. 7.

[43] NIST Special Publication 800-53, 2013, p. 23.

[44] NIST Special Publication 800-37, 2014, p. 7.

substantially formulated or developed before entering the security risk management process. The system must be sufficiently defined to be categorized. At this point, it is likely already too late to significantly influence the security architecture. Thus, security requirements under DoD and the NIST RMF are defined, in a sense, backward: The requirements are defined by the predetermined solutions that meet them, rather than defining mission-, outcome-, or performance-based requirements from which solutions are derived through system engineering and design.

The spirit of the NIST RMF is to secure commercial, off-the-shelf information systems procured by the government for which design, architectures, protocols, and interfaces are already substantially determined. The security controls were defined with the mindset of what would need to be applied to such systems. The controls do not explicitly address in any depth how design, architecture, protocols, and interfaces should be shaped to achieve acceptable cybersecurity. As argued in Chapter One, these choices are critical for cybersecurity, especially in determining how resilient and robust a military system is to attack. And, in general, measures incorporated in design, architecture, protocols, and interfaces afford better anti-access measures than security controls. For military systems (including NSS, PIT, and PIT systems), the government can control design, architecture, protocols, and interfaces and need not rely solely on established security controls chosen and applied afterward as overlays.

Despite the flexibility to control design in NSS, the CNSS, chaired by the DoD CIO, further defines cybersecurity policy for NSS by adopting (with minor modifications) the security categorizations and controls, supplemented by the development of standardized "overlays," which are sets of controls and guidance tailored for particular types of NSS.[45] This specification, together with the DoD RMF, severely restricts DoD's freedom to manage cybersecurity of PIT systems or other NSS differently from how it manages general information technology systems via security controls imposed after system design.

In May 2014, NIST released an initial public draft of a publication on system security engineering that moves in the direction of integrating cybersecurity considerations throughout the entire system life cycle, including the initial design phase, so that "system security requirements and security concept of operations provide the basis for the architectural design, implementation, verification, and validation of the system."[46] However, this guidance was written after the NIST RMF policy and is not integrated into it; hence the risk management processes outlined in the various documents do not align with one other and are not incorporated in the DoD RMF.[47]

---

[45] See CNSS Instruction 1253, 2012, for discussion and guidance on application of overlays.

[46] Ross, Oren, and McEvilley, 2014, p. 44.

[47] For example, NIST Special Publication 800-160 (Ross, Oren, and McEvilley, 2014) specifies "planning for risk management" (p. 25) as the first risk management step, while NIST Special Publication 800-39 begins with "framing risk" (p. 33).

Cybersecurity and Air Force Life-Cycle Management

As outlined above, the bulk of cybersecurity policy in DoD is structured around the management of individual *systems*, rather than mission-level security.[48] These policies are the responsibility of the agency heads and CIO. On the other hand, programming and acquisition are structured *programs*, and, as will be developed later in this chapter, many of the cybersecurity processes use program structures such as acquisition milestones for implementation. Programs do not necessarily map cleanly to systems: A system may grow out of materiel managed under multiple programs, while conversely a single program may contain multiple discrete systems. Further, there are legacy systems that do not fall under the purview of a program office. DoD cybersecurity policy specifies that a system manager be assigned for each such system, with the same responsibilities for cybersecurity that a program manager has for the system(s) within a program.[49] For convenience in the discussion that follows, we generally refer only to the program manager, and we will refer to systems and programs interchangeably, except where a specific distinction is made.

In this section, we describe how cybersecurity is managed throughout the life cycle of a military system in the Air Force. The description is current as of August 2014 and is drawn dominantly from formal policy, but we also include some observations on the real implementation of policy we discovered during unstructured discussions with relevant stakeholders.

Roles and Responsibilities

Authorizing Officials

Many of the cybersecurity responsibilities of the Secretary of the Air Force and CIO that are mandated by law and DoD policy are carried out under the DoD RMF by the authorizing officials. Authorizing officials determine what level of risk to accept for systems containing information technology based on a balance of concerns from mission and cybersecurity perspectives.

Authorizing officials grant or deny several forms of authorizations. For a system to be operated in the field, an authorization to operate (ATO) is required, granted after the authorizing official determines that the overall cybersecurity risk is acceptable. The authorizing official may also, with permission from the component CIO, determine that there are noncompliant controls with too high residual risk, but that the overall risk is acceptable given the mission criticality, and grant a conditional ATO. Such an ATO must be reviewed after a year to ensure that the risks

---

[48] The DoD RMF policy requiring Tier 1 and Tier 2 risk management is very new and has not been fully implemented through the Air Force or other services. As such, it will take time to determine the efficacy of the new policies in improving the overall state of cybersecurity and mission assurance.

[49] DoD Instruction 8510.01, 2014, p. 10.

have been reduced.[50] In some cases, a full ATO might be required for test "if operational testing and evaluation is being conducted in the operational environment or on deployed capabilities," but sometimes an interim authority to test is adequate.[51] An ATO, once granted, has a termination date within three years of issue, meaning that it is reviewed on a calendar-driven basis rather than an event-driven basis, such as a change in the threat environment. However, if an approved continuous monitoring program is in place, there is no termination date.[52]

If the authorizing official assesses that the residual risk has not been reduced to an acceptable level, a denial of authorization to operate can be issued. From our discussions, this seems to be a rare occurrence: More typically, the authorizing official will negotiate with the program manager on mitigation actions and delay a decision rather than issue a denial of authorization to operate.

Authorizing officials perform these tasks as independent assessors, responsible to the component CIO when performing authorizing official duties. The goal is that all systems containing information technology fall under an authorizing official. From our discussions with authorizing officials in the Air Force, we find that the assignment of authorizing officials to particular systems is a reactive process. When systems require an authorization, the program or system manager seeks the approval of an authorizing official. For most systems and programs, the appropriate authorizing official is clear. For those for which the appropriate assignment is ambiguous, Air Force practice is that one authorizing official will choose to take on the system. In cases where no authorizing official accepts the system, the assignment of the system defaults to the Air Force senior information security officer.

Because of the large number of programs in sustainment that predate current cybersecurity policies and that have limited funds or program support, as of 2014, there are many Air Force systems that do not have an assigned authorizing official. For older systems without an explicit ATO, it is unclear what mechanism, short of a major modification, would bring them to the attention of an authorizing official or make operators aware that the system needs cybersecurity review.

DoD component heads and CIOs are responsible for appointing authorizing officials. Policy specifies that authorizing officials "should be appointed from senior leadership positions within business owner and mission owner organizations . . . to promote accountability in authorization decisions that balance mission and business needs and security concerns," but "the AO [authorizing official] cannot be or report to the [program manager]/[system manager] or program executive officer."[53] In practice, this means that authorizing officials are not typically part of the CIO staff but reside instead in other organizations (both Air Staff and major commands) for organize, train, and equip responsibilities.

---

[50] DoD Instruction 8510.01, 2014, p. 25.

[51] DoD Instruction 8510.01, 2014, p. 36.

[52] DoD Instruction 8510.01, 2014, p. 35.

[53] DoD Instruction 8510.01, 2014, pp. 17, 29.

Authorizing officials are selected from senior leaders in these organizations but are not provided with dedicated staff for the purpose of cybersecurity risk acceptance. They are supported by security control assessors, who provide expert assessment and advice, but authorizing officials do not appoint or directly supervise these individuals. The Air Force has more than a dozen authorizing officials, and many that oversee military systems reside in Air Force Materiel Command. This organizational position also means that the authorizing officials do not have budgetary authority over the systems they authorize.

Program Offices

The principal means by which cybersecurity authorizations are initiated is through the acquisition process, which means that systems not under program management receive less scrutiny. Managing system cybersecurity vulnerability for a program is the responsibility of the program manager and is managed and monitored through the program protection plan, required for all programs.[54] The program protection plan is a general plan for security protection of a program; it can be thought of as the plan for keeping "secret things from getting out" and "keeping malicious things from getting in."[55] In addition to a cybersecurity plan mandated for all systems containing information technology, the program protection plan also includes a security classification guide, a counterintelligence support plan, a criticality analysis, and an anti-tamper plan. The program protection plan is reviewed by the milestone decision authority at each of the milestone decisions and the full-rate production decision review, but the cybersecurity plan must also be approved by the appropriate authorizing official. The program protection plan should be reviewed, and updated if necessary, at least once every three years.[56]

The mandatory cybersecurity strategy, included as an appendix to the program protection plan, is approved at all milestone decisions and contract awards.[57] The program protection plan, including the cybersecurity plan, is built around the concepts of critical program information and mission-critical functions and components, which are results of the criticality analysis.[58] A program's list of critical program information and critical components is required to be reviewed periodically—every three years, upon receipt of updated threat assessments (which are requested every three years); if there are significant changes to the program; and if monitoring identifies countermeasure effectiveness degradation.[59] However, it is unclear whether a cyber-related incident would be included in the countermeasure monitoring, which appears to be focused on

---

[54] DoD Instruction (Interim) 5000.02, 2013, p. 84.

[55] Defense Acquisition University, 2013, Section 13.0.

[56] Air Force Pamphlet 63-113, 2013, p. 31.

[57] DoD Instruction (Interim) 5000.02, 2013, p. 132.

[58] Critical program information is defined in DoD Instruction 5200.39, 2010, p. 17. The process for identifying mission-critical functions and components is outlined in Defense Acquisition University, 2013, Section 13.3.2.

[59] Air Force Pamphlet 63-113, 2013, pp. 30–31.

program protection surveys at contractor facilities and the loss or theft of critical program information and critical components.[60]

Although policy directs that identifying vulnerabilities should in theory encompass all aspects of a program, it "should begin with these critical functions, associated components and [critical program information]."[61] As a result, systems that contain components or information not deemed critical can receive less cybersecurity scrutiny, particularly when resources are scarce. As noted in Chapter One, the critical vulnerabilities for cyber access are not always the most critical components, which can lead to lack of attention to vulnerabilities of a system that reside in components that are not deemed critical.

Identification of software vulnerabilities in critical components is performed through a variety of software tools that can identify common vulnerabilities, rather than mandating or evaluating software for secure system design and good programming practices.[62] Another enumerated software vulnerability is the amount of access given to third parties during software development.[63]

The program protection plan includes a risk assessment step that combines the threat and vulnerability analyses of critical program information and critical components to evaluate the overall risk to the system from existing vulnerabilities and assessed threats. This risk assessment process identifies appropriate countermeasures that include anti-tamper, information assurance (focused on critical program information), software assurance (focused on security controls), supply chain risk management, trusted suppliers, and system security engineering.[64] Although system security engineering is listed as a possible countermeasure, it is presented as more of an overlay on the entire program protection plan. The risk assessment described here is separate from the RMF process for cybersecurity that is partitioned into the cybersecurity strategy appendix. It is important to note that the program protection plan as a whole is reviewed by the milestone decision authority, but the cybersecurity strategy appendix is independently reviewed and approved through the DoD CIO chain.

The primary emphasis of the program protection plan is on the design and acquisition phases of a system life cycle, rather than operations and sustainment. According to policy, the program protection plan should be validated for currency and technical content prior to transition to operations and sustainment, and any major modification requires an update or a new program protection plan.[65] Also, "repair depots should be aware of critical program information . . . and

---

[60] Air Force Pamphlet 63-113, 2013, pp. 60–62.

[61] Defense Acquisition University, 2013, Section 13.5.

[62] DoD Instruction (Interim) 5000.02, 2013, p. 85; Defense Acquisition University, 2013, Section 13.5.1.

[63] Defense Acquisition University, 2013, Section 13.5.1.

[64] Defense Acquisition University, 2013, Section 13.6.

[65] Air Force Pamphlet 63-113, 2013, p. 31.

mission-critical functions and components on systems they are maintaining so as to appropriately protect these items from compromise."[66]

For programs with an up-to-date program protection plan and cybersecurity strategy, monitoring of security controls is performed on an annual basis, rather than being event-driven and affected by the report of a particular cyber-related incident.[67] However, not all programs have these components: Legacy programs that have been in sustainment for many years were never involved in this process, and it is unclear how they might be monitored or updated to reflect the current policy.

Mission Owners

The mission owner—called the information system owner in cybersecurity policy—for a military system plays three key roles regarding cybersecurity: (1) establishing the cybersecurity requirements for mission success (including mission impact of cybersecurity failures), (2) training for and operating the system in the field, and (3) participating in the programming and budgeting process that funds cybersecurity solutions for military systems, especially those in the sustainment phase of the life cycle.

The primary process for establishing cybersecurity requirements according to mission impact is the Joint Capabilities Integration and Development System (JCIDS). According to cybersecurity policies, evaluating mission impact is accomplished by "categorizing" the information system according to a process outlined in policy. The authorizing official has the ability to participate in the categorization process.[68] The categorization involves assigning "impact values (low, moderate, or high) reflecting the potential impact on organizations or individuals should a security breach occur (i.e., a loss of confidentiality, integrity, or availability)."[69]

The policies acknowledge that evaluating mission impact of any given system requires visibility across all the systems that participate in the mission, putting this activity at Tier 2 of the risk management process.[70] Note that this description assumes an already-defined system that fits into an existing "enterprise architecture." This framework will likely be typical for the acquisition of business systems and network components, but it does not address the fact that military systems can be designed in multiple ways, each of which may result in different sets of security categorizations.

---

[66] Defense Acquisition University, 2013, Section 13.10.4.

[67] DoD Instruction 8510.01, 2014, p. 30.

[68] NIST Special Publication 800-37, 2014, p. 21.

[69] CNSS Instruction 1253, 2009, p. 4.

[70] NIST SP 800-37, 2014, p. 21.

Once the system is categorized, the categorization is documented in the "appropriate JCIDS capabilities document (e.g., capabilities development document)."[71] This process is the primary way in which cybersecurity requirements are captured in a program. The capabilities development document also includes key performance parameters that capture the essential functionality of the system, but there is no mandatory key performance parameter for cybersecurity. In our discussions with program offices, we repeatedly and consistently heard that the information system owners specified no requirements for cybersecurity beyond system categorization. As a result, cybersecurity becomes purely an exercise in applying security controls focused on information technology systems, which is consistent with stated policy.

The military system owner also has responsibilities for training and operating the systems, and these activities contribute to the cybersecurity of the systems. As noted in Chapter One, the security of a system is a combination of the system itself and how humans use it. Vulnerabilities can come from either the system design or how it is used, and hence mitigations can also be found in both design and use. Outside of prohibition of certain actions, such as the use of universal serial bus (USB) devices, little policy directs how cybersecurity should be achieved on the operational side. The information system owner is directed to assign a user representative for each information system and PIT system, who is "responsible for ensuring that user operational interests are met throughout the systems authorization process."[72] This user representative could offer potential mitigations from the user side and play a role in a process that finds acceptable solutions from a balance of design and use considerations. However, beyond this very general statement of responsibilities, the user representative is given no specific responsibilities or authorities in the DoD RMF.

Finally, the military system owner plays a key role in programming and budgeting for cybersecurity of military systems. In this role, the largest challenge is the fiscally constrained environment, but we focus here on policy matters, and the chief constraint on the policy side is that programming is structured around programs and not systems. This structure can hinder the allocation of resources for cybersecurity for military systems not under program control, such as numerous small systems in the sustainment phase that are funded out of general operations and sustainment accounts. Even though these systems are small, they can nonetheless be the Achilles' heel for cyber attack.

Supporting Functions, Monitoring, and Feedback

The two main supporting functions for acquisition and life-cycle management that provide monitoring and feedback on cybersecurity are intelligence (including counterintelligence) and test.

---

[71] DoD Instruction 8510.01, 2014, p. 18.

[72] CNSS Instruction 4009, 2010, p. 80.

Intelligence

Cybersecurity intelligence support to acquisition is one part of general intelligence support to acquisition. Intelligence support to acquisition was undergoing a review in 2014, with the goal of large changes; this discussion describes the state of support as of August 2014. The full extent of issues that present challenges for intelligence support is beyond the scope of this report.

The principal intelligence products provided to program offices come in the form of formal reports issued at specific junctures in the acquisition process. Capstone Threat Assessments are performed in support of JCIDS and treat a general threat area rather than a program. System Threat Assessment Reports are produced for major acquisition programs at milestone A and are required again at milestone C.[73] They determine the relevant operational threat environment used during operational testing.[74]

However, threat assessments are only regulatory for major defense acquisition programs and major automated information system programs—smaller programs will likely not receive the tailored threat assessments needed to perform a rigorous risk analysis.[75] Specific assessments can also be requested by a program office regarding supply chain risks. These assessments are performed by the Defense Intelligence Agency Threat Assessment Center, but the overall capacity of that organization is limited, and not all programs and systems get assessed.[76]

For programs in sustainment, updated threat assessments for critical program information and critical components are supposed to be requested every three years.[77] Note that this is a calendar-rather than event-driven approach and does not generally allow for updating the threat analysis based on an emerging threat or newly documented incident and is only based on previously identified critical program information or critical components.

Classification issues, resource constraints, and the difficulty of translating specific cyber incident reports into language and recommendations relevant to affected systems limit cybersecurity feedback from the intelligence community to programs and mission owners. Intelligence on cyber threats can be highly classified and involve U.S. persons information or law enforcement sensitive information, limiting access to working-level members of program offices and those in charge of system security.

Specific intelligence products relevant to a program flow directly to the program office. Other intelligence, such as cyber security breaches and other incidents, are monitored by acquisition intelligence personnel and channeled to program offices as deemed relevant. We could find no policy that directs that such intelligence be sent to relevant authorizing officials.

---

[73] DoD Instruction (Interim) 5000.02, 2013, pp. 52, 57.

[74] DoD Instruction (Interim) 5000.02, 2013, p. 94.

[75] DoD Instruction (Interim) 5000.02, 2013, p. 57.

[76] Air Force Pamphlet 63-113, 2013, p. 6; Defense Acquisition University, 2013, Section 13.4.1.

[77] Air Force Pamphlet 63-113, 2013, p. 29.

The acquisition intelligence must support a broad array of programs with limited resources and a limited understanding of how those programs are connected. In conversations with a variety of actors throughout the acquisition process, we learned that in many cases, they feel that the intelligence reports they receive are of limited value for improving their system's cybersecurity. Systems in the sustainment phase present a particularly difficult problem for feedback from the intelligence community, as in many cases the program office is very resource-constrained or may not exist at all.

Test

Test provides the most direct monitoring and feedback on performance and security. Military system programs typically require testing through every phase of their life cycle, excepting disposal. The purpose of test and evaluation (T&E) is to verify for all stakeholders that the system or component under test meets technical and performance specifications (developmental test) or to verify that it is operationally effective and suitable for use in its intended environment (operational test). T&E is the primary mechanism for obtaining information about the actual state of cybersecurity within a system (including, but not limited to, verifying compliance with security controls) and passing that information to the relevant stakeholders, including the authorizing official.

The primary customer for developmental test and evaluation (DT&E) is the program manager. The program manager's primary objective for T&E is for it to assess program development progress and to identify problems as early in the program's life cycle as possible to minimize the cost and schedule impacts of redesign and rework. Results are also used to support authorizing decisions of the authorizing official.[78]

Program test planning is documented in a Test and Evaluation Master Plan (TEMP), which includes developmental and operational test requirements, schedules, funding, resources, organizational responsibilities, and testing methodologies.[79] To better integrate development test activities and resources, as well as those needed for operational test, program managers establish an integrated test team. The integrated test team is composed of representatives from the program office, operational testers, the developmental test activities, and the system contractor, as well as the acquisition, requirements, operations, intelligence, and sustainment communities, as appropriate. Every program has a designated Lead Developmental Test and Evaluation Organization that supports test planning and analysis, leads government developmental test execution, and coordinates with any participating test organizations as needed.[80]

---

[78] The coordination of T&E requirements with those for cybersecurity certification and authorization is currently done by the integrated test team and documented in the system TEMP.

[79] Defense Acquisition University, 2013, Section 9.0; Air Force Instruction 99-103, 2013, pp. 55–56.

[80] Defense Acquisition University, 2013, Section 9.0; Air Force Instruction 99-103, 2013, pp. 55–56.

Whereas DT&E is focused on testing the technical characteristics of a system, operational test and evaluation (OT&E) is concerned with assessing whether or not DoD personnel can operate the system effectively in likely operational scenarios.[81] Performance is assessed in two broad areas: operational effectiveness[82] and operational suitability, guided by the measures of effectiveness documented in the Initial Capabilities Document and Capabilities Development Document.[83] System survivability and operational security are also evaluated as major contributors to effectiveness and suitability.[84] Specific critical operational issues to be evaluated in operational testing are documented in the TEMP.

The Air Force Operational Test and Evaluation Center (AFOTEC) commander is responsible to the Chief of Staff of the Air Force for the independent OT&E of all major and nonmajor system acquisitions. AFOTEC reviews operational requirements, employment concepts, tactics, maintenance concepts, and training requirements and provides early operational assessments during technology development and operational assessments during engineering and manufacturing development to inform decisionmaking in addition to its primary mission of conducting formal OT&E on production representative systems. In addition to conducting testing of systems in acquisition, AFOTEC also supports information assurance and interoperability assessments of operational combatant command systems. Once AFOTEC-led testing is completed, the major commands assume responsibility for conducting any cybersecurity testing of fielded systems.

A variety of approaches may be used in cybersecurity testing depending on the objectives of the test, the nature of the system to be tested, the scope of the test, and the time and resources available to execute the testing. The methodologies are categorized by NIST as

- review
- target identification and analysis
- target vulnerability validation.[85]

---

[81] See 10 U.S.C. § 139 and 2366.

[82] "DoD defines *operational effectiveness* as the overall degree of mission accomplishment of a system when used by representative personnel in the environment planned or expected for operational employment of the system considering organization, training, doctrine, tactics, survivability or operational security, vulnerability, and threat" (Defense Acquisition University, 2013, Section 9.3.2.1; italics added).

[83] *Operational suitability* defines the degree in which a system satisfactorily places in field use, with consideration given to reliability, availability, compatibility, transportability, interoperability, wartime usage rates, maintainability, safety, human factors, manpower supportability, logistics supportability, documentation, environmental effects, and training requirements (Defense Acquisition University, 2013, Section 9.3.2.1).

[84] "Survivability or operational security includes the elements of susceptibility, vulnerability, and recoverability. As such, survivability or operational security acts as an important contributor to operational effectiveness and suitability. All systems under OT&E oversight should receive survivability or operational security assessment if exposed to threat weapons in a combat environment or to combat-induced conditions that may degrade capabilities, regardless of designation for LFT&E [Live Fire Test & Evaluation] oversight." (Defense Acquisition University, 2013, Section 9.3.2.3)

[85] NIST Special Publication 800-115, 2008, pp. 2-2, 2-3.

The test teams that perform dedicated cybersecurity testing are often referred to as "blue" or "red" teams. Near the end of system development, a blue team can be brought in to work with program office personnel to examine system architecture, design, interfaces, security controls, and operating procedures, to identify potential vulnerabilities. This information is provided to the program manager to develop a prioritized list of deficiencies to correct prior to initial OT&E. It is also used to support authorizing official decisions.

For cybersecurity operational testing, an independent operational test team, acting as a blue team, first reviews system documentation, security controls and the plan of action, and milestones for correcting cybersecurity deficiencies. The team uses a variety of tools to develop its own vulnerability and penetration assessment in an operationally representative environment. During a subsequent operational test event, a red team attempts to penetrate and exploit the system during normal operations. The red team simulates a cyber attack and therefore operates covertly, using only the knowledge that could be realistically gained by a dedicated attacker.[86] Robust red team testing thus provides the most realistic assessment of all aspects of system cybersecurity posture against real-world threats, including network defenses. This is also the only testing that realistically assesses robustness and resilience in the face of cyber attack, since it tests not only the system but operating procedures and operator response as well. Such testing is, however, often limited by time and the availability of suitable testing environments and personnel.[87]

Specialized infrastructure is required to support cybersecurity testing. The Air Force Materiel Command has created an Information Technology Test Range consisting of the Air Operation Center and Datalinks Test Facilities at Eglin Air Force Base, the Capabilities Integration Environment at Maxwell-Gunter Air Force Base, and the C4ISR Enterprise Integration Facility at Hanscom Air Force Base. The DoD Test Resource Management Center sponsors the Joint Mission Environment Test Capability, which provides infrastructure and protocols for a wide range of system and network testing. Although these facilities are well suited for cybersecurity testing of business and command, control, communications, computers, intelligence, surveillance, and reconnaissance (C4ISR) systems and their networks, they are less capable over the broad range of PIT systems. As a result, PIT system cybersecurity testing often has to compete for time and access to system integration laboratories or operational platforms to provide appropriate test environments. In addition, due to the specialized nature of PIT systems, there are few automated data reduction and analysis tools available to improve test team productivity.

---

[86] Director of Operational Test and Evaluation Memorandum, 2014, p. 3.

[87] Red team testing generally requires a secure environment and measures to prevent unintended disruption or compromise of real-world systems and activities. Operational test planning attempts to balance the need for testing in a realistic operational environment, which includes operating with interconnected systems, and limiting adverse effects on operations not part of the system under test.

A highly qualified and experienced staff is needed to design and execute cost-effective test programs tailored to the needs of each system and utilize the capabilities of this sophisticated equipment. For this reason, T&E personnel tend to be specialized in both their technical discipline and the type of testing on which they focus. This expertise is developed by combining technically qualified personnel and hands-on experience in planning, executing, and analyzing tests. The availability of qualified cyber personnel for testing is a significant limitation and is particularly acute in the case of red teams. The Air Force currently has only two certified red teams, one of which is in the Air National Guard and is in the process of losing billets occupied by experienced full-time cybersecurity personnel. Unlike the Army and Navy, the Air Force has no red teams with acquisition program testing as a primary mission. As a result, when Air Force teams were not available, AFOTEC has had to request red team support from other services or the Defense Information Systems Agency, whose expertise is more with information systems than weapon systems.

Continuous Monitoring

FISMA requires that government agencies file annual reports on the state of cybersecurity across the enterprise.[88] Preparing these reports has proven to be time-consuming and expensive, and the results not particularly timely or valuable.[89] In response, the Office of Management and Budget mandated a switch to a continuous monitoring system in April 2010, with the Department of Homeland Security taking the lead on implementation. The department has developed a Continuous Diagnostics and Mitigation (CDM) program, which[90]

> enables Federal Government departments and agencies to expand their continuous diagnostic capabilities by increasing their network sensor capacity, automating sensor collections, and prioritizing risk alerts. CDM offers commercial off-the-shelf (COTS) tools, with robust terms for technical modernization as threats change. First, agency-installed sensors perform an automated search for known cyber flaws. Results feed into a local dashboard that produces customized reports, alerting network managers to their worst and most critical cyber risks, based on standardized and weighted risk scores. Prioritized alerts enable agencies to efficiently allocate resources based on the severity of the risk. Progress reports track results, which can be shared among sister networks. Summary information can feed into an enterprise-level dashboard to inform and prioritize cyber risk assessments.[91]

The recognition that continuous monitoring is critical for the fast-evolving threats of cybersecurity is a welcome one; however, the CDM program, as so much of cybersecurity activity in DoD, is focused on developing solutions for business systems and networks and does

---

[88] 44 U.S.C. § 3544.

[89] Spoth, 2010.

[90] Spoth, 2010, and U.S. Office of Management and Budget Memorandum M-10-15, 2010, pp. 1–2.

[91] U.S. Department of Homeland Security, 2014.

not reflect assessed priorities for weapon systems. The standardized tools and sensors described here are unlikely to represent appropriate tools for the wide range of military systems that also require cybersecurity monitoring.

Authorities and Accountability

The various actors in cybersecurity have some overlapping authorities, resulting in ambiguous accountability for cybersecurity of military systems. Authorizing officials, through issuing ATOs, "are accountable for the security risks associated with information system operations."[92] Policy places authorizing officials in a position to determine whether a system is allowed to operate or connect, but they do not control resources and therefore cannot determine trades between cybersecurity and other critical program or system parameters. Those decisions are allocated to the program manager, who is ultimately responsible for the program (and system) in the form of ensuring that cost, schedule, and performance (including cybersecurity) requirements are met. Further, although the authorizing official determines the level of cybersecurity risk to be assumed, so does a military commander. A military commander in an operational setting has the authority to make decisions on what risk to assume, and these decisions can come in conflict with directions from authorizing officials.

Conclusion

The stakeholders for cybersecurity in the Air Force are confronted with a welter of laws and policies that are voluminous, complicated, and changing faster than the life cycle of a military system. The spirit of the governance is motivated by the problem of securing information technology systems that are largely commercial, off-the-shelf more so than military systems for which the government has some control over design, architecture, protocols, and interfaces. The governance also prescribes how cybersecurity is to be achieved, and that solution is largely via security controls applied to a system as an overlay. System security engineering does not play a prominent role. Because the implementation of much of cybersecurity policy is done at acquisition milestones, programs are emphasized more than systems, programs in procurement are emphasized more than those in sustainment, and large programs have more oversight than small programs. Monitoring and feedback on exactly how well all this is done is incomplete and weak.

This imposed governance structure places considerable bounds on the Air Force's options for cybersecurity—not just on what to do, but how it is to be done. Nevertheless, the Air Force has opportunities within these constraints to shape a cybersecurity policy that achieves more effective performance. The next chapter first summarizes shortfalls of the governance described

---

[92] NIST Special Publication 800-39, 2011, p. D-4.

in this chapter relative to the ideals described in Chapter One, then offers a way forward for Air Force policy to address these shortcomings.

3. Findings and Recommendations

> "The challenge of course for all involved in government and industry is how to ensure that prophylactic measures do not end up creating worse symptoms than the diseases we are seeking to avert. It is possible to have too much of the wrong sort of security."
>
> – David Omand [1]

A comparison of the processes and oversight for cybersecurity in life-cycle management described in the previous chapter with the principles described in Chapter One indicates a number of shortcomings in the current Air Force management of cybersecurity. In this chapter, we discuss these shortcomings and their consequences and recommend several courses of action for partial redress. We examined dimensions of cybersecurity across all stakeholders, but our emphasis was on those aspects under the control of the Air Force and the life-cycle management community. We discerned numerous insights at many levels of hierarchy in the Air Force across many functions. Those highlighted here focus on the higher levels of policy—how to shape the overall framework of cybersecurity in the Air Force. Addressing many of these issues by our recommendations should facilitate more detailed solutions at lower levels.

Resource and time constraints limited the findings and recommendations we were able to reach in some supporting areas. For example, there are clear limitations to the capacity of the test organizations to provide the full scope of feedback that the life-cycle management community and authorizing officials could beneficially use, but we were not able to fully document the current capacities and demands. There are clear shortfalls in the relationship between the intelligence and acquisition communities, and those shortfalls extend far beyond cybersecurity. Assessing those shortfalls is a larger problem than we were able to address, and the relationship between these two communities was under review as this project was being executed. Also, vulnerabilities exist in the defense industrial base. A lesser emphasis on these supporting issues should not be construed as meaning that we assess their state to be satisfactory.

We summarize in the next section four key findings that we assess to be root causes of deficiencies in cybersecurity management of military systems in the Air Force. For each finding, we also list a number of associated consequences that hinder cybersecurity and mission assurance.

---

[1] Omand, 2010, p. 75.

Findings

Finding #1

The cybersecurity environment is complex, rapidly changing, and difficult to predict, but the policies governing cybersecurity are better suited to simple, stable, and predictable environments, leading to significant gaps in cybersecurity management.

Enterprises that operate in simple, stable, and predictable environments often adopt an organizational design that includes centralized control executed through standardized and formalized work rules. As the primary means for achieving cybersecurity, DoD has recently standardized and formalized that the DoD components are to manage cybersecurity of all systems via specific NIST security controls implemented through the RMF. But cybersecurity is a complex task, the threat and technologies are rapidly changing, and the future threats are not easy to predict. Such a list of security controls, even with tailored overlays, is unlikely to anticipate and address all possible vulnerabilities in these complex systems, remain current as the threats and technologies change, and anticipate all future issues. Given that the adversary needs to find only one vulnerability to exploit and the Air Force needs to operate through any plausible attack, this management choice is likely to fail in the long run. The focus on security controls has several additional problematic consequences.

Consequence #1. It prescribes solutions for military system cybersecurity in the form of controls that are not as comprehensive in providing security as sound system security engineering. Controls overlaid on top of a sound system architecture and design can be effective; it is unlikely that any set of controls can mitigate issues that arise from a fundamentally flawed architecture or system design. And controls primarily address the ability to access a system, not the robustness or resiliency of that system once breached. That DoD instructs the use of security controls does not prohibit program offices from adopting sound system security engineering. But the construction of policy oversight of cybersecurity around such controls to meet program protection plan requirements, especially in a resource-constrained environment, consumes resources and thereby does not create incentives to pursue more comprehensive system security engineering measures that improve robustness and resiliency.

Consequence #2. The processes and security controls were developed principally with information technology systems in mind, not military systems. The list of controls can be tailored to a military system, but the resulting tailored list is not the soundest approach possible to dealing with military systems. The very different characteristics of information technology systems and military systems point to the need for effective system security engineering early in the acquisition process that is catered to the requisite cybersecurity needs of each military system.

Consequence #3. The strategic goal of mission assurance is diminished in favor of tactical security controls. As complete cybersecurity is unlikely to ever be attainable, a sensible approach for managing cybersecurity is risk management. This approach is adopted in the Risk

Management Framework being implemented throughout DoD. Yet, the implementation of that framework is written with a strong emphasis on security controls, thereby focusing on the tactical Tier 3 and diminishing the central role of *mission assurance* in Tier 2 of the RMF.[2] The RMF does provide for perspectives that embrace what is called mission assurance in the DoD, but the emphasis on security controls makes it possible to meet the requirements of that framework with little consideration of mission assurance.

Consequence #4. Relying on standardized and formalized security controls as the basis for cybersecurity, the policy telegraphs to the enterprise that the implicit goal of cybersecurity is compliance with security controls. This message is not the intended message—the real goal of cybersecurity is to achieve clear outcomes, such as keeping the operational impact of adversary cyber exploitation and offensive cyber operations to an acceptable level in light of mission assurance.

Finding #2

The implementation of cybersecurity is not continuously vigilant throughout the life cycle of a military system, but instead is triggered by acquisition events, mostly during procurement, resulting in incomplete coverage of cybersecurity issues by policy.

Oversight of cybersecurity occurs mainly when the program protection plan is written and revised and when developmental and operational test events occur and are assessed. All acquisitions of systems containing information technology, including national security systems, are required to have a cybersecurity strategy, approved by the CIO reporting chain, as part of their program protection plan.[3] This strategy is to be updated at major milestones and at the full rate production decision. The developmental and operational tests also transpire at specific junctures in the acquisition process. Both the program protection plan and the various test events are discrete events concentrated in the procurement phase of a military system. There are four principal consequences of this intermittent triggering of attention to cybersecurity and the timing of those triggers.

Consequence #1. These programmatic events come late in the design process, and therefore provide little leverage to influence some critical design decisions that affect cybersecurity. As early as the research and development phase, sometimes even before a project formally becomes a program, decisions on design, architecture, interfaces, and protocols, which can significantly shape cybersecurity, are already firm. The *first step* of the NIST RMF is to "categorize the information system,"[4] which presumes the system already is defined enough for categorization, and is therefore late in the design process. If these early design decisions are made without an adequate balancing of functionality and security, an inherently insecure design can be the

---

[2] See DoD Instruction 8510.01, 2014.

[3] DoD Instruction (Interim) 5000.02, 2013.

[4] NIST Special Publication 800-37, 2010, p. 21.

outcome, and security controls superimposed on the system later will struggle to compensate for the deeper vulnerabilities introduced by the design choices.

Consequence #2. Systems in programs beyond procurement, being sustained or disposed, get diminished attention relative to those in procurement. Although procurement is a critical phase because design decisions are made that shape the cybersecurity of the system, systems are vulnerable throughout their life cycles, especially in light of the evolving threat. The culture among authorizing officials, especially given the other demands on them, is that their responsibility is to decide whether to accept risk when asked to initially certify a system during the acquisition process or to renew certification for a system in sustainment. This policy incentive is quite different from giving an authorizing official a specific portfolio of projects, systems, or programs for which that official has continuous responsibility. The RMF now instructs that an authorizing official be appointed in writing for "all DoD [information systems] and [PIT] systems operating within or on behalf of the DoD Component," so this issue might be mitigated by this policy, but the current practice falls short of this directive, and the policy seems to give latitude of what might be considered a relevant system and does not include initiatives in the project stage, before they become programs.[5]

Because of this intermittent triggering of cybersecurity oversight weighted toward procurement, legacy systems and programs beyond the procurement phase get less scrutiny, and some (many?) have no visibility to the authorizing officials at all. Just because these programs might be small or beyond procurement does not at all imply that they are not important for effective cybersecurity. If one of these is the weak point in a connected system of systems, it might be the most important vulnerability. Further, a number of modifications are made to systems that do not rise to a level that requires a revised program protection plan or authorizing official review. These include some sustainment modifications and modifications that incur little cost (and might be paid for out of the user's operational account), but can potentially alter the cybersecurity of the system.[6]

Consequence #3. This policy structure leads to imbalanced risk assessment by favoring system vulnerability assessments over mission impact and threat. The overall mission risk is a combination of the vulnerabilities that the supporting system possesses (that might change when the system is modified), the threat posed by the adversaries, and the consequences to supported missions. By linking cybersecurity oversight triggers to acquisition events, changes of concern are examined only when the system is procured or major modifications are made, not when the threat environment for a system changes or when some other change occurs—say, to concepts of operation or changes to other systems—that might affect the consequences to the operational

---

[5] DoD Instruction 8510.01, 2014, Enclosure 2, Section 7(c).

[6] Some of these are called "1067 modifications" after the form used for these modifications. See Air Force Instruction 63-131, 2013.

mission. These latter changes can occur between calendar-driven reviews for programs beyond procurement.

Consequence #4. Management, oversight, and budgeting within DoD are strongly structured around programs, whereas cybersecurity vulnerabilities cross program boundaries. Programmatically structured oversight fails to embrace the fundamental cross-system and cross-program nature of cybersecurity.

Finding #3

Control of and accountability for military system cybersecurity is spread over numerous organizations and is poorly integrated, resulting in diminished accountability and diminished unity of command and control for cybersecurity.

Many stakeholders play important roles in cybersecurity. The main ones are (1) the using command(s) for a military system, (2) the acquisition/life-cycle management and test communities, (3) the authorizing officials and CIO reporting chain, and (4) the intelligence and counterintelligence communities. There is nothing inherently wrong with spreading the accountability and control over numerous organizations. Indeed, the diversity of tasks in cybersecurity is consistent with assigning them to these multiple stakeholders. When multiple stakeholders are involved in a single process, however, blurring of roles and responsibilities diminishes control and accountability relative to when roles and responsibilities are very clearly established. This clarity is lacking in some areas of cybersecurity.

The existence of the authorizing official reporting to a CIO is somewhat duplicative of other authorities and diminishes accountability. The authorizing official decides what level of cybersecurity risk to accept in a program. But the operational command that employs a military system also has authorities to accept operational risk, including cybersecurity risks. And, only that commander can assess risks to mission assurance, given that the full scope of considerations needed to make that assessment lies within that commander's ambit. The authorizing official also presumably holds accountability for the decisions he or she makes, although we cannot find anywhere in policy that specifically holds the authorizing official accountable. But, the program manager is clearly accountable for the performance of systems in his or her program, including cybersecurity. These overlapping roles create ambiguity of who can make the final decisions regarding risk to a mission—the commander or the authorizing official? And, should a cybersecurity incident occur, who is ultimately to be held accountable—the program manager, the authorizing official, or the operational commander?

Finding #4

Monitoring and feedback for cybersecurity is incomplete, uncoordinated, and insufficient for effective decisionmaking or accountability.

Monitoring and feedback for cybersecurity comes in several forms. During the acquisition process, signals are sent by authorizing officials to the acquisition chain and CIO regarding *compliance with security controls.* Following a risk assessment framework, these officials assess which controls to implement and whether the program office carries out these controls. Without testing, however, authorizing officials have limited insight into the effectiveness of these controls, or even into whether the controls were in reality put on contract and fielded as designed.

Feedback on effectiveness of cybersecurity measures comes from developmental and operational test and evaluation on *the meeting of requirements and the performance of the systems in simulated operational environments.* This information gets back to the program office, but not by any formal mechanism to the authorizing official. Although we were not able to fully probe the capacities and constraints on the test side, it seems likely that this feedback is not comprehensive enough in scope to discover anywhere near all vulnerabilities, and insufficient authorities appear to exist to explore robustness and resiliency. An annual report by the director of OT&E gives feedback to Congress on the performance of military systems, including cybersecurity, but this feedback is limited to major defense acquisition programs and selected other programs chosen at the director's discretion. Most testing happens early in a program, and programs in sustainment receive little testing unless they undergo modifications. Systems outside program oversight are subject to little to no testing. Hence, testing provides limited feedback on the cybersecurity of legacy military systems.

A third feedback signal comes from the intelligence and counterintelligence communities, who collect, analyze, and disseminate information regarding *threats and actual incidents.* The demand signal from the acquisition community to the intelligence community for what kinds of information are most useful to them is weak, and information specific to military systems is not always finding its way back to key stakeholders in acquisition and the authorizing officials.

There is little cross-flow of information among these three feedback streams. The consequences of the nature and the sparseness of the monitoring and feedback are threefold.

Consequence #1. There are several critical gaps in this feedback:

1. None of these monitoring and feedback mechanisms systematically surveys all systems or programs (including those in sustainment and very small programs), and therefore the effort lacks sufficient scope.
2. None goes far into probing the operational consequences of any cybersecurity shortfalls, and therefore feedback does not properly inform risk management of mission assurance.
3. The various filaments of feedback from the independent sources are not combined in any way that is digestible and useable for a program office or authorizing official for decisionmaking or for holding any decisionmakers accountable.

Consequence #2. The lack of comprehensive, program-oriented (or system-oriented) feedback on cybersecurity and the impact (real or potential) of cybersecurity on operational missions is in stark contrast to the abundance of feedback on cost and schedule performance. This imbalance in feedback creates an incentive structure for program managers and program executive officers that favors emphasis on cost and schedule over performance, specifically cybersecurity.

Consequence #3. These deficiencies in feedback on cybersecurity inhibit individual accountability. Like safety, cybersecurity is everyone's responsibility in an organization. To hold individuals accountable in a fair manner throughout an organization, the feedback on their role in any cybersecurity problem needs to be accurate and timely. Such feedback is currently lacking, and those who make mistakes or violate policies do not appear to be held to the same account as those who violate other orders. If an airman opens an attachment to email that contains malware, or makes an unauthorized connection of a computer to the Internet, the monitoring and consequent disciplinary actions do not appear to be as tight as occurs if an airman violates a technical order or places an unauthorized part in an aircraft.

Discussion

In the above list of findings, some will probably find it surprising that a need for clearer cybersecurity requirements is omitted. In our discussions with stakeholders across the Air Force and DoD, the need for cybersecurity requirements was raised repeatedly. The lack of clear requirements for cybersecurity does in fact make it harder to fund cybersecurity solutions that compete for cost, schedule, and performance with other priorities. The more performance objectives can be precisely expressed in key performance parameters and other specifications, the more they compete as priorities throughout the acquisition process.

Yet it is not clear to us or to any with whom we spoke what form this language should take. The reason the form of this language is perplexing derives from the basic characteristics of cybersecurity we describe in Chapter One—specifically that cybersecurity is highly technical, decisions on cybersecurity are intertwined with the functionality of systems and how systems are used, the threats in cyberspace are rapidly evolving, and decisions regarding cybersecurity should be made in the context of mission assurance. These characteristics mean that cybersecurity solutions need to be tailored to individual systems, must be adaptive and evolve with the threat, and must be highly integrated with the design of the systems and the mission(s) those systems support. These characteristics are particularly the case for denying access. Requirements that are specific enough to be placed on contract and used as benchmarks for operational testing are unlikely to be sufficient. Specific and effective cybersecurity solutions are more likely to arise from sound system security engineering.

Cybersecurity is more than denying access—it also comprises robust and resilient designs. Robustness and resiliency are intrinsic attributes of a military system and are like other survivability requirements. Requirements for robustness and resiliency mandate system characteristics such as diversity and redundancy in order to maintain a specified level of functionality when subsystems fail, regardless of why they fail. These kinds of requirements can be put on contract in the same way that robust power delivery in an aircraft can be put on contract—by requiring diverse and redundant power systems. Such requirements for cybersecurity might include a specified level of functionality in the face of events such as the loss of data integrity, failure of a processor, failure of a communication link, or failure of an entire computer system.

For these reasons, we assess—with caveats—that changes to the key performance parameters that drive a prioritization for cybersecurity would be welcome if they elevate cybersecurity so that it competes with comparable concerns in the acquisition and programming process trade-space decisions. But this language would need to direct that designs lend themselves to be adaptable to meet desired cybersecurity outcomes *as the threat evolves*, not just satisfy static requirements for denial of access. Requirements for robustness and resiliency could be incorporated into the survivability key performance parameter. A parallel example is survivability in the context of electronic warfare. The desired performance outcome in that case is that the system is modifiable in an affordable manner in the face of changing threats, not that designed systems have firm solutions that foresee all future conceivable electronic warfare threats.

These changes could take the form of making cybersecurity a part of the mandatory force protection or survivability key performance parameters, or establishing cybersecurity as a separate mandatory key performance parameter. But the dilemma is that if this language is too vague, the requirement has the potential to have little impact, and if the language is too specific, it is likely to do more harm than good by imposing specific guides (like security controls) that can be satisfied but are insufficient in providing effective cybersecurity.

Recommendations

These four root causes, which we have listed in the form of findings and their resultant consequences, are structurally imbedded in DoD. No simple solution exists that will correct them all. Some result from well-intentioned statutory requirements and DoD policies that are not easily changed. But some room exists within these bounds for the Air Force to adjust policies to redress, at least partially, many of these problems. A theme throughout the recommendations discussed in this section is that more needs to be done in many areas of cybersecurity, and this naturally means that more resources will probably be required.

Recommendation #1

Define cybersecurity goals for military systems within the Air Force around desired outcomes, while remaining consistent with DoD issuances.

In a hierarchy of objectives moving from strategies at the top to tasks at the bottom, the highest-echelon objective is the goal the enterprise sets out to accomplish. This goal has not been clearly articulated for cybersecurity in DoD. We recommend a succinct, *outcome-oriented* objective for cybersecurity that applies to all systems throughout their life cycles, regardless of the degree to which they contain "information technology." The working definition we use is *to keep the impact of adversary cyber exploitation and offensive cyber operations to an acceptable level as guided by a standardized process for assessing risk to mission assurance.* This kind of direction would focus actors in the Air Force on outcomes, regardless of the program or other activities in which they are involved. It would also foster a pervasive culture within the Air Force that every member has a role in cybersecurity, much like the message to all airmen that they all play a role in safety. We recommend that the Air Force define desired cybersecurity outcomes along these lines in headquarters-level Air Force issuances. (Addresses Finding #1, Fourth Consequence; because the lack of outcome-based goals inhibits outcome-based feedback, partially addresses Finding #4; facilitates Recommendations #9, #10, #11, and #12.)

Recommendation #2

Realign functional roles and responsibilities for cybersecurity risk assessment around a balance of system vulnerability, threat, and operational mission impact and empower the authorizing official to integrate and adjudicate among stakeholders.

Current management structures for cybersecurity do not integrate and balance the three components of risk: (1) vulnerabilities of systems, (2) threats to those systems, and (3) the impact to operational missions. No formal mechanisms exist for channeling all the relevant intelligence and test data to all the stakeholders, especially to the authorizing officials. And, no formal mechanisms exist to adjudicate competing cybersecurity solutions between the program office and the using command, such as when a solution might be found through design (but have high cost) or change in how the system is used (but might be less foolproof).

To address these shortcomings, we propose a framework for functional roles and responsibilities that is explicitly constructed around a balance of these three components of risk, depicted in Figure 3.1.[7] The life-cycle management community, specifically the program manager, would be responsible for program and system vulnerabilities and would advocate for mitigation in the context of program cost, schedule, and performance. The intelligence and counterintelligence communities would be responsible for threat assessments. The CFLI or other mission owner would be responsible for operational mission assurance and advocate from this perspective.

---

[7] We drew inspiration for this approach from the frameworks proposed in Lewis, Coggin, and Roll, 1994; and Tripp et al., 2006.

Figure 3.1. Framework for Cybersecurity Management Roles and Responsibilities

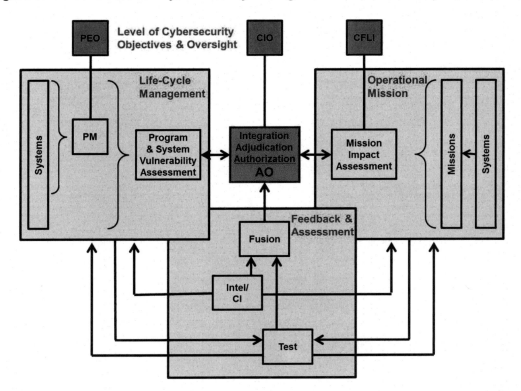

NOTE: AO = authorizing official; CI = counterintelligence; PEO = program executive officer; and PM = program manager.

Each of the three components of risk would have a single assessor that advocates on behalf of his or her corresponding component. The assessor for vulnerabilities of systems and programs could be the security control assessor in the new RMF. To be most useful for oversight, an integrator in the Air Force could combine threat assessments from the intelligence community with real observations of cybersecurity performance gleaned through testing and red team results, but, at a minimum, policy could direct that existing tailored threat information of the highest feasible classification level and all test reports flow to the relevant authorizing official. The mission impact could be assessed on the mission owner side by the user representative in the new RMF. The authorizing officials would then integrate these three inputs of risk, adjudicate competing proposed solutions, and perform a cybersecurity risk acceptance and authorization balanced across vulnerability, threat, and impact.

The advantages of this construct over the current one are that functional roles and responsibilities would be clearly delineated by the three components of risk, and a single individual, the authorizing official, would be empowered to integrate and adjudicate issues across the three components and the key stakeholders. The disadvantages are that additional resources would be required to expand the role of the user representative to assess mission impact on a comparable level with the security control assessor, whose role would also need to expand to be a program and system vulnerability assessor. We also acknowledge that standing up an organization to fuse feedback and assessments might not be actionable given resource constraints, but might be justified if it can support other stakeholders beyond the authorizing officials. At minimum, we recommend that (1) some single point of contact for missions be assigned within the mission owners to advocate on behalf of mission assurance to the authorizing officials, (2) the authorizing officials be explicitly given integration and adjudication duties, and (3) all relevant intelligence and test reports that go to a program office also flow to the relevant authorizing official. (Partially addresses Finding #3; facilitates Recommendations #8, #11, and #12.)

Recommendation #3

Assign authorizing officials a portfolio of systems and ensure that all systems comprehensively fall under some authorizing official throughout their life cycles.

Systems throughout the Air Force are not currently comprehensively assigned to authorizing officials. As mentioned above under Finding #2, the intent of the DoD RMF is that an authorizing official be appointed in writing for "all DoD [information systems] and [PIT] systems operating within or on behalf of the DoD Component."[8] Satisfactory compliance with that direction might be satisfied by attempting to identify all relevant systems in the Air Force and assigning each to an authorizing official. Such a bottom-up approach risks not being comprehensive, either because systems are overlooked or because systems are not deemed relevant. Further, such an approach would not capture initiatives early in the design phase when they are still projects, prior to becoming formal programs. Ideally, the authorizing officials would be involved as the system design decisions are being made. To close this gap more assuredly, we recommend that authorizing officials be assigned a specific portfolio for which they are responsible, including identifying any new relevant systems, even at the project stage. We further recommend that the authorizing officials have a defined role in the Initial Capabilities Document and the Analysis of Alternatives Study Guidance in order to participate in shaping design considerations that affect cybersecurity. These changes would reverse their role from being reactive (addressing authorization requests when solicited) to proactive (maintaining continuous oversight of a portfolio). To meet these responsibilities, authorizing officials would need additional resources. (Partially addresses Finding #2, First, Second, Third, and Fourth Consequences.)

---

[8] DoD Instruction 8510.01, 2014, Enclosure 2, Section 7(c).

Recommendation #4

Adopt, within the Air Force, policy that encourages program offices to supplement the required security controls with more comprehensive cybersecurity measures, including sound system security engineering.

Security controls are central to the DoD implementation of the RMF, but we see no impediment to the Air Force supplementing this framework with instructions to program offices and authorizing officials to proactively use additional Air Force defined instructions to assess the adequacy of system security engineering. Security controls at best address only denial of access to systems and do not address robust or resilient design. Security controls also do not address access vulnerabilities that are introduced due to poor design, such as choice of architecture, protocols, and interfaces. We recommend adopting a broader view of cybersecurity by making the authorizing officials the champions for system security engineering in the Air Force. Doing so would require new Air Force policy and additional resources to carry out these additional responsibilities. (Partially addresses Finding #1, First and Second Consequences; facilitates Recommendation #5.)

Recommendation #5

Foster innovation and adaptation in cybersecurity by decentralizing in any new Air Force policy how system security engineering is implemented within individual programs.

We recommend that decisions on objectives and strategies for cybersecurity be centralized and the decisions on how to carry out those objectives and strategies be decentralized. The expertise for identifying cybersecurity problems and solutions resides throughout the enterprise, but is concentrated in the engineering disciplines and program offices. Organizational designs that centralize these decisions and standardize and formalize the tasks for achieving cybersecurity increase control and efficiency, but diminish innovation and adaptation. Decentralizing these decisions is how organizations typically cope with complex, rapidly changing, and unpredictable environments such as cybersecurity. Attempting to standardize and formalize how system security engineering is implemented, for example, in the way that controls are used to standardize and formalize the RMF, would reduce this process to a similar compliance exercise and suppress innovation.

In addition, a free exchange of information should be encouraged among workers to share information and potential solutions. A corollary to this recommendation is that all members of the acquisition community need to be educated in the basics of cybersecurity, much like all airmen are inculcated in the idea of safety. Therefore, we recommend implementing the system security engineering solutions in Recommendation #4 without imposing a new set of standardized and formalized controls, but rather by devolving this responsibility to the program offices and authorizing officials. (Partially addresses Finding #1.)

Recommendation #6

Reduce the overall complexity of the cybersecurity problem by explicitly assessing the cybersecurity risk/functional benefit trade-off for all interconnections of military systems in cyberspace, thereby reducing the number of interconnections by reversing the default culture of connecting systems whenever possible.

The most actionable way to reduce complexity is to strengthen the review of what is allowed to be connected to what in cyberspace, both by design and in practice in the field. Air Force culture, especially for military systems, should shift from a default of connecting military systems to one another and to information technology networks to one that requires an explicit assessment of each proposed interface. This assessment should evaluate the balance between cybersecurity consequences (and consequences to operational risk) and the operational capability the interface would enable. This assessment would be done by authorizing officials on both sides of the interface. For example, if an aircraft is proposed to be connected to a piece of maintenance equipment, both the authorizing official for the aircraft and for the maintenance equipment should authorize the connection. Further, program offices should develop a culture of striving for simplicity in design in order to reduce complexity. (Partially addresses Finding #1.)

Recommendation #7

Create a group of experts in cybersecurity that can be matrixed as needed within the life-cycle community, making resources available to small programs and those in sustainment.

Create an organization within the AFLCMC or its directorates that can provide a range of skills to program offices, including system security engineers and experts in offensive and defensive cyber operations. The core talent in this organization would reside at the working level and assist in anticipating vulnerabilities and rapidly finding innovative solutions. This organization would be matrixed to support programs across the center as needed, regardless of program size or phase of life cycle. The existence of this talent pool would not preclude large programs from having their own dedicated staffs. This organization could report to the AFLCMC/EN (Engineering Directorate), or each program executive officer could maintain such an organization catered to his or her portfolio. (Partially addresses Finding #2, Second Consequence.)

Recommendation #8

Establish an enterprise-directed prioritization for assessing and addressing cybersecurity issues in legacy systems.

Mitigating all the issues in legacy systems will require time and money. Given limited resources, not all the cybersecurity issues in legacy systems can be identified and mitigated immediately. Effort in addressing identified issues needs to be focused on the most critical first. The methodology for establishing what is critical in this prioritization should be guided by what mitigates vulnerabilities in *operational missions* first, and which operational missions, rather than a prioritization of *systems*. This prioritization could be facilitated by the functional construct outlined in Recommendation #2, where a user representative assesses mission impacts and an authorizing official uses this information to direct a prioritization to the appropriate program executive officers and program managers. (Partially addresses Finding #2, Second Consequence.)

Recommendation #9

Close feedback gaps and increase visibility of cybersecurity by producing a regular, continuous assessment summarizing the state of cybersecurity for every program in the Air Force and hold program managers accountable for a response to issues.

We recommend closing these gaps in two ways: (1) Increase the overall quantity of assessments that are performed, especially those that reflect the real state of cybersecurity (as opposed to compliance with security controls)—test data and red team results—and (2) ensure that all the relevant information flows to the key stakeholders, especially the authorizing officials. The flow of feedback on cybersecurity outcomes is incomplete, and no formal mechanism exists to pass information on cybersecurity outcomes to the authorizing officials. As authorizing officials have no staff, we recommend having some kind of organization in the Air Force, perhaps in AFLCMC, collate and edit relevant intelligence and test information (including red team results) that pertains to systems under their purview. It is important that this feedback not be constrained to quantifiable metrics, but that it capture the assessed state of cybersecurity, in qualitative terms when necessary. Preparing this assessment will require a range of technical knowledge and familiarity with intelligence assessments. We recommend that this assessment accompany reports on program performance that include cost and schedule and be classified at the appropriate level when necessary. We recognize that this recommendation is resource-intensive in a resource-constrained environment. If resources do not permit such a fusion, at the very minimum, we recommend that intelligence reports and test reports that are sent to a program office also be copied to the relevant authorizing official. (Partially addresses Finding #2, Third Consequence; addresses Finding #4, First and Second Consequences; facilitates Recommendation #11.)

Recommendation #10

Create cybersecurity red teams within the Air Force that are dedicated to acquisition/life-cycle management.

The Air Force has only two certified red teams, and neither has acquisition program support as its primary mission. The competing other duties of these teams restricts their availability for acquisition. The Army and the Navy possess red teams dedicated to cybersecurity in acquisition. Creation of dedicated teams made up of experts providing continuity of support (e.g., civilian or reserve) would increase the ability to monitor the true state of cybersecurity. (Addresses Finding

#1, First and Second Consequences; partially addresses Finding #4, First and Second Consequences; facilitates Recommendation #11.)

Recommendation #11

Hold individuals accountable for infractions of cybersecurity policies.

When policies dictate certain actions or the prohibition of certain actions, appropriate consequences should follow if violated. We recommend that the increased monitoring and feedback from Recommendation #9 be used to hold individuals accountable when possible. Deliberate infractions could go on the security records of personnel; punishments need not be heavy for occasional violations, but repeat offenses should bring meaningful consequences, perhaps up to and including revocation of security clearances. Implementing this recommendation will require building over time meaningful mechanisms for monitoring behavior and determining appropriate disciplinary actions. (Addresses Finding #4, Third Consequence.)

Recommendation #12

Develop mission thread data to support program managers and authorizing officials in assessing acceptable risks to missions caused by cybersecurity deficiencies in systems and programs.

Mission assurance results from the interaction of multiple systems and how they are used. Some situational awareness of where individual systems within a program office's responsibilities contribute to a mission requires an understanding of the mission threads. Mission owners should take the lead in defining mission threads to assist in making informed decisions on mission risk and communicate these in a way that assists the authorizing officials in accepting cybersecurity risk. This role could be performed by the RMF's user representative via the construct in Recommendation #2. (Partially addresses Finding #1, Third Consequence, and Finding #2, Third Consequence.)

Closing Remarks

Adopting all of these recommendations will not result in perfect cybersecurity. Nevertheless, following these recommendations should better position the Air Force to manage cybersecurity risk by improving situational awareness and providing sharper means for mitigating the threat. But these recommendations cannot be implemented effectively in isolation. Some of the recommendations depend on other improvements to work. The lack of a clearly articulated desired outcome for cybersecurity leaves individual actors within the Air Force without clear direction. Lack of monitoring and feedback leaves decisionmakers at all levels without requisite information to make critical decisions or to hold individuals accountable. Lack of clearly defined roles and responsibilities inhibits the integration and coordination of individuals.

Many of the recommendations will also require increased resource allocation for personnel, infrastructure, and education and training. No changes to policies will be effective without an adequately educated and trained workforce to implement them. As emphasized in Chapter One, cybersecurity is a highly technical field that changes rapidly, and effective cybersecurity measures must be integrated into the very designs of military systems via system security engineering. Like safety, cybersecurity requires the vigilance of all members of the Air Force. They all need some understanding of this domain. Some individuals need to be champions of system security engineering in the Air Force, and these officials need to have deep and up-to-date knowledge of engineering and the threat. If cybersecurity is to be more than an application of security controls to an already designed system, program offices need to be populated with personnel who understand both system security engineering and the mindset and tactics of adversaries determined to attack through cyberspace. All airmen who operate and maintain military systems need to understand how their actions can expose a well-designed system to risk and how to operate it securely in a contested cyberspace environment.

Our hope is that if the Air Force adopts these recommendations, it will result in a significant step toward mitigating the advanced threat through cyberspace.

Abbreviations

AFLCMC	Air Force Life Cycle Management Center
AFOTEC	Air Force Operational Test and Evaluation Center
ATO	authorization to operate
CA	certification authority
CDM	Continuous Diagnostics and Mitigation
CFLI	Core Function Lead Integrator
CIO	chief information officer
CNSS	Committee on National Security Systems
DAA	designated approving (or accrediting) authority
DIACAP	DoD Information Assurance Certification and Accreditation Process
DITSCAP	DoD Information Technology Security Certification and Accreditation Process
DoD	U.S. Department of Defense
DOTMLPF-P	doctrine, organization, training, materiel, leadership and education, personnel, and facilities, together with policy
DT&E	developmental test and evaluation
FISMA	Federal Information Security Management Act of 2002
JCIDS	Joint Capabilities Integration and Development System
NIST	National Institute of Standards and Technology
NSS	national security systems
OT&E	operational test and evaluation
PAF	RAND Project AIR FORCE
PIT	platform information technology
RMF	Risk Management Framework
T&E	test and evaluation
TEMP	Test and Evaluation Master Plan
USB	universal serial bus
U.S.C.	U.S. Code

References

Air Force Instruction 63-131, *Modification Management*, Washington, D.C.: Secretary of the Air Force, March 19, 2013.

Air Force Instruction 91-204, *Safety Investigations Reports*, Washington, D.C.: Secretary of the Air Force, February 12, 2014 (corrective actions applied on April 10, 2014).

Air Force Instruction 99-103, *Capabilities-Based Test and Evaluation*, Washington, D.C.: Secretary of the Air Force, October 16, 2013.

Air Force Pamphlet 63-113, *Program Protection Planning for Life Cycle Management*, Washington D.C.: Secretary of the Air Force, October 17, 2013.

Anderson, Ross, *Security Engineering*, 2nd ed., Indianapolis, Ind.: Wiley, 2008.

Baldwin, Kristen, Judith Dahmann, and Jonathan Goodnight, "Systems of Systems Security: A Defense Perspective," *Insight*, Vol. 14, No. 2, 2011, pp. 11–13. As of February 23, 2015: http://www.acq.osd.mil/se/docs/SoS-Security-INCOSE-INSIGHT-vol14-issue2.pdf

Bayuk, Jennifer L., Dennis Barnabe, Jonathan Goodnight, Drew Hamilton, Barry Horowitz, Clifford Neuman, and Stas' Tarchalski, *Systems Security Engineering: A Research Roadmap*, Systems Engineering Research Center, Report No. SERC-2010-TR-005, August 22, 2010.

Bayuk, Jennifer L., Jason Healey, Paul Rohmeyer, Marcus H. Sachs, Jeffrey Schmidt, and Joseph Weiss, *Cyber Security Policy Guidebook*, Hoboken, N.J.: John Wiley & Sons, 2012.

Bayuk, Jennifer L., and Barry M. Horowitz, "An Architectural Systems Engineering Methodology for Addressing Cyber Security," *Systems Engineering*, Vol. 14, No. 3, 2011, pp. 294–304.

CNSS—*See* Committee on National Security Systems.

Committee on National Security Systems Instruction 1253, *Security Characterization and Control Selection for National Security Systems*, March 15, 2012.

Committee on National Security Systems Instruction 4009, *National Information Assurance (IA) Glossary*, April 26, 2010.

Committee on National Security Systems Policy 22, *Policy on Information Assurance Risk Management for National Security Systems*, January 2012.

Damanpour, Fariborz, "Organizational Innovation: A Meta-Analysis of Effects of Determinants and Moderators," *The Academy of Management Journal*, Vol. 34, No. 3, September 1991, pp. 555–590.

Defense Acquisition University, *Defense Acquisition Guidebook*, incorporating changes through September 2013.

Department of Defense Directive 5200.28, *Security Requirements for Automatic Data Processing (ADP) Systems*, Washington D.C.: Under Secretary of Defense for Acquisition, December 18, 1972, cancelled March 21, 1988.

Department of Defense Directive 5200.28, *Security Requirements for Automatic Information Systems (AISs)*, Washington, D.C.: Under Secretary of Defense for Acquisition, March 21, 1988, cancelled October 24, 2002.

Department of Defense Directive 8500.01E, *Information Assurance (IA)*, Washington, D.C.: DoD Chief Information Officer, April 23, 2007.

Department of Defense Handbook MIL-HDBK-1785, *System Security Engineering Program Management Requirements*, Washington, D.C.: U.S. Department of Defense, August 1, 1995.

Department of Defense Instruction (Interim) 5000.02, *Operation of the Defense Acquisition System*, Washington, D.C.: U.S. Department of Defense, November 26, 2013.

Department of Defense Instruction 5200.39, *Critical Program Information (CPI) Protection Within the Department of Defense*, Washington D.C.: Under Secretary of Defense for Intelligence, July 16, 2008, incorporating Change 1, December 28, 2010.

Department of Defense Instruction 5200.40, *DoD Information Technology Security Certification and Accreditation Process (DITSCAP)*, Washington D.C.: Assistant Secretary of Defense for Command, Control, Communications, and Intelligence, December 30, 1997, cancelled November 28, 2007.

Department of Defense Instruction 8500.01, *Cybersecurity*, Washington, D.C.: DoD Chief Information Officer, March 14, 2014.

Department of Defense Instruction 8510.01, *DoD Information Assurance Certification and Accreditation Process (DIACAP)*, Washington, D.C.: Department of Defense Chief Information Officer, November 28, 2007.

Department of Defense Instruction 8510.01, *Risk Management Framework (RMF) for DoD Information Technology (IT)*, Washington, D.C.: Department of Defense Chief Information Officer, March 12, 2014.

Director of Operational Test and Evaluation Memorandum, *Procedures for Operational Test and Evaluation of Cybersecurity in Acquisition Programs*, August 1, 2014.

Fischer, Eric A., *Federal Laws Relating to Cybersecurity: Overview and Discussion of Proposed Revisions*, Congressional Research Service Report R42114, June 20, 2013.

Gosler, James R., and Lewis Von Thaer, *Resilient Military Systems and the Advanced Cyber Threat*, Washington, D.C.: Department of Defense Defense Science Board, January 2013.

Jensen, Michael C., and William H. Meckling, "Specific and General Knowledge, and Organizational Structure," in L. Werin and H. Hijkander, eds., *Contract Economics*, Cambridge, Mass.: Basil Blackwell, 1992, pp. 251–274.

Joint Publication 1-02, *Department of Defense Dictionary of Military and Associated Terms*, Washington, D.C.: Joint Chiefs of Staff, November 8, 2010 (as amended through July 16, 2014). As of February 23, 2015:
http://www.dtic.mil/doctrine/new_pubs/jp1_02.pdf

Kahneman, Daniel, *Thinking, Fast and Slow*, New York: Farrar, Straus and Giroux, 2011.

Kendall, Frank, and Daniel M. Tangherlini, *Improving Cybersecurity and Resilience Through Acquisition*, Washington, D.C.: Department of Defense and General Services Administration, November 2013.

Krekel, Bryan, Patton Adams, and George Bakos, *Occupying the Information High Ground: Chinese Capabilities for Computer Network Operations and Cyber Espionage*, Northrop Grumman Corporation, March 7, 2012.

Lewis, Leslie, James A. Coggin, and Charles Robert Roll, Jr., *The United States Special Operations Command Resource Management Process: An Application of the Strategy-to-Tasks Framework*, Santa Monica, Calif.: RAND Corporation, MR-445-A/SOCOM, 1994. As of February 23, 2015:
http://www.rand.org/pubs/monograph_reports/MR445.html

Manual for the Operation of the Joint Capabilities Integration and Development System, January 19, 2012.

Maybury, Mark T., *Cyber Vision 2025: United States Air Force Cyberspace Science and Technology Vision, 2012–2025*, Washington, D.C.: United States Air Force Chief Scientist, December 13, 2012.

Mihm, Jürgen, Christoph H. Loch, Dennis Wilkinson, and Bernardo A. Huberman, "Hierarchical Structure and Search in Complex Organizations," *Management Science*, Vol. 56, No. 5, May 2010, pp. 831–848.

Mintzberg, Henry, *The Structuring of Organizations: A Synthesis of the Research*, Englewood Cliffs, N.J.: Prentice-Hall, 1979.

National Institute of Standards and Technology, *Standards for Security Categorization of Federal Information and Information Systems*, Federal Information Processing Standards Publication 199, February 2004.

————, *Minimum Security Requirements for Federal Information and Information Systems*, Federal Information Processing Standards Publication 200, March 2006.

National Institute of Standards and Technology Special Publication 800-37, *Guide for Applying the Risk Management Framework to Federal Information Systems: A Security Life Cycle Approach*, Revision 1, February 2010, including updates as of June 5, 2014.

National Institute of Standards and Technology Special Publication 800-39, *Managing Information Security Risk: Organization, Mission, and Information System View*, March 2011.

National Institute of Standards and Technology Special Publication 800-53, *Recommended Security Controls for Federal Information Systems and Organizations*, Revision 3, August 2009.

National Institute of Standards and Technology Special Publication 800-53, *Security and Privacy Controls for Federal Information Systems and Organizations*, Revision 4, April 2013.

National Institute of Standards and Technology Special Publication 800-115, *Technical Guide to Information Security Testing and Assessment*, September 2008.

NIST—*See* National Institute of Standards and Technology.

Omand, David, *Securing the State*, New York: Columbia University Press, 2010.

Public Law 96-511, *Paperwork Reduction Act of 1980*, December 11, 1980.

Public Law 104-106, *National Defense Authorization Act for Fiscal Year 1996*, February 10, 1996.

Public Law 107-347, *E-Government Act of 2002*, December 17, 2002.

Reason, James, *Human Error*, New York: Cambridge University Press, 1990.

Reynard, W. D., C. E. Billings, E. S. Cheaney, and R. Hardy, *The Development of the NASA Aviation Safety Reporting System*, NASA Reference Publication 1114, November 1986.

Roche, James G., and Barry D. Watts, "Choosing Analytic Measures," *Journal of Strategic Studies*, Vol. 14, 1991, pp. 165–209.

Ross, Ron, Janet Carrier Oren, and Michael McEvilley, *Systems Security Engineering: An Integrated Approach to Building Trustworthy Resilient Systems*, NIST Special Publication 800-160 (Initial Public Draft), May 2014.

Scott, W. Richard, *Organizations: Rational, Natural, and Open Systems*, 3rd ed., Englewood Cliffs, N.J.: Prentice Hall, 1992.

Snyder, Don, Bernard Fox, Kristin F. Lynch, Raymond E. Conley, John A. Ausink, Laura Werber, William Shelton, Sarah A. Nowak, Michael R. Thirtle, and Albert A. Robbert, *Assessment of the Air Force Materiel Command Reorganization: Report for Congress*, Santa Monica, Calif.: RAND Corporation, RR-389-AF, 2013. As of March 26, 2015:
http://www.rand.org/pubs/research_reports/RR389.html

Spoth, Tom, "OMB Issues New Rules on IT Security," *Federal Times*, May 2, 2010. As of February 23, 2015:
http://www.federaltimes.com/article/20100502/IT01/5020306/
OMB-issues-new-rules-security

Steinbruner, John D., *The Cybernetic Theory of Decision*, Princeton, N.J.: Princeton University Press, 1974.

Tripp, Robert S., Kristin F. Lynch, Charles Robert Roll, Jr., John G. Drew, and Patrick Mills, *A Framework for Enhancing Airlift Planning and Execution Capabilities Within the Joint Expeditionary Movement System*, Santa Monica, Calif.: RAND Corporation, MG-377-AF, 2006. As of March 26, 2015:
http://www.rand.org/pubs/monographs/MG377.html

U.S. Code, Title 10, Armed Forces, Subtitle A, General Military Law, Part I, Organization and General Military Powers, Chapter 4, Office of the Secretary of Defense, Section 139, Director of Operational Test and Evaluation. As of March 26, 2015:
http://www.gpo.gov/fdsys/granule/USCODE-2011-title10/
USCODE-2011-title10-subtitleA-partI-chap4-sec139

U.S. Code, Title 10, Armed Forces, Subtitle A, General Military Law, Part IV, Service, Supply, and Procurement, Chapter 139, Research and Development, Section 2366, Major Systems and Munitions Programs: Survivability Testing and Lethality Testing Required Before Full-Scale Production. As of March 26, 2015:
http://www.gpo.gov/fdsys/granule/USCODE-2011-title10/
USCODE-2011-title10-subtitleA-partIV-chap139-sec2366

U.S. Code, Title 15, Commerce and Trade, Chapter 7, National Institute of Standards and Technology, Section 278g-3, Computer Standards Program. As of March 26, 2015:
http://www.gpo.gov/fdsys/granule/USCODE-2009-title15/
USCODE-2009-title15-chap7-sec278g-3

U.S. Code, Title 40, Public Buildings, Property, and Works, Subtitle III, Information Technology Management, Chapter 111, General, Section 11101, Definitions. As of March 26, 2015:
http://www.gpo.gov/fdsys/granule/USCODE-2011-title40/
USCODE-2011-title40-subtitleIII-chap111-sec11101

U.S. Code, Title 44, Public Printing and Documents, Chapter 35, Coordination of Federal Information Policy, Subchapter II, Information Security, Section 3532, Definitions. As of March 26, 2015:
http://www.gpo.gov/fdsys/granule/USCODE-2011-title44/USCODE-2011-title44-chap35-subchapII-sec3532

U.S. Code, Title 44, Public Printing and Documents, Chapter 35, Coordination of Federal Information Policy, Subchapter III, Information Security, Section 3544, Federal Agency Responsibilities. As of March 26, 2015:
http://www.gpo.gov/fdsys/granule/USCODE-2008-title44/USCODE-2008-title44-chap35-subchapIII-sec3544

U.S. Department of Defense, *Department of Defense Mission Assurance Strategy*, April 2012.

U.S. Department of Homeland Security, "Continuous Diagnostics and Mitigation (CDM)," 2014. As of March 26, 2015:
http://www.dhs.gov/cdm

Vicente, Kim, *The Human Factor*, New York: Routledge, 2003.